Mathematics Research Developments

nova
science publishers
www.novapublishers.com

Mathematics Research Developments

Frontiers in Mathematical Modelling Research
M. Haider Ali Biswas and M. Humayun Kabir (Editors)
2022. ISBN: 978-1-68507-430-2 (Hardcover)
2022. ISBN: 978-1-68507-845-4 (eBook)

Mathematical Modeling of the Learning Curve and Its Practical Applications
Charles Ira Abramson and Igor Stepanov (Authors)
2022. ISBN: 978-1-68507-737-2 (Hardcover)
2022. ISBN: 978-1-68507-851-5 (eBook)

Partial Differential Equations: Theory, Numerical Methods and Ill-Posed Problems
Michael V. Klibanov and Jingzhi Li (Authors)
2022. ISBN: 978-1-68507-592-7 (Hardcover)
2022. ISBN: 978-1-68507-727-3 (eBook)

Outliers: Detection and Analysis
Apra Lipi, Kishan Kumar, and Soubhik Chakraborty (Authors)
2022. ISBN: 978-1-68507-554-5 (Softcover)
2022. ISBN: 978-1-68507-587-3 (eBook)

Advanced Engineering Mathematics and Analysis. Volume 2
Rami A. Maher (Author)
2022. ISBN: 978-1-68507-605-4 (Hardcover)
2022. ISBN: 978-1-68507-691-7 (eBook)

Advanced Engineering Mathematics and Analysis. Volume 1
Rami A. Maher (Author)
2021. ISBN: 978-1-53619-869-0 (Hardcover)
2021. ISBN: 978-1-68507-435-7 (eBook)

More information about this series can be found at
https://novapublishers.com/product-category/series/mathematics-research-developments/

Gennadiy Zhizhin

Non-Euclidean Geometry in Materials of Living and Non-Living Matter in the Space of the Highest Dimension

nova
science publishers
www.novapublishers.com

DOI: https://doi.org/10.52305/NVUE5435

Library of Congress Cataloging-in-Publication Data

Names: Zhizhin, G. V. (Gennadiĭ Vladimirovich), author.
Title: Non-Euclidean geometry in materials of living and non-living matter in the space of the highest dimension / Zhizhin Gennadiy Vladimirovich.
Description: New York : Nova Science Publishers, [2022] | Series: Mathematics Research Developments | Includes bibliographical references and index. |
Identifiers: LCCN 2022027553 (print) | LCCN 2022027554 (ebook) | ISBN 9781685078850 (hardcover) | ISBN 9798886970647 (adobe pdf)
Subjects: LCSH: Polytopes. | Dimension theory (Topology) | Biomolecules--Mathematical models. | Geometrical models.
Classification: LCC QA691 .Z46 2022 (print) | LCC QA691 (ebook) | DDC 516/.158--dc23/eng20220826
LC record available at https://lccn.loc.gov/2022027553
LC ebook record available at https://lccn.loc.gov/2022027554

Published by Nova Science Publishers, Inc. † New York

Contents

Preface vii

Chapter 1 **Euclidean Geometry and Non-Euclidean Geometry** 1

Chapter 2 **Geometry of the Polytopes of Higher Dimensions** 9

Chapter 3 **Non-Euclidean Properties of the Geometry of Polytopes of Higher Dimension** 39

Chapter 4 **Polytopes of the Highest Dimension of Inert Substances** 69

Chapter 5 **"Inert" Substances as a Self-Regulating Medium Tending to Capture Space** 103

Chapter 6 **Spatial Models of Sugars and Their Compounds** 127

Chapter 7 **The Theory of the Folder and Native Structures of the Proteins** 159

Chapter 8 **Geometry of the Structure of Nucleic Acids in the Space of the Highest Dimension** 189

Chapter 9 **Interaction of Nucleic Acids in the Space of Higher Dimension and the Transmission of Hereditary Information** 219

Chapter 10 **Dimension of Substances and Life** 253

Summary 265

References 267

About the Author 277

Index 279

Preface

One of the most important scientists in the field of natural science, V.I. Vernadsky, starting with research in mineralogy (Vernadsky, 1917 - 1921), who created the doctrine of the biosphere, introduced the concept of living and inert matter (Vernadsky, 1926). "The living matter of the biosphere is the totality of living organisms living in it. The bulk of the substances of the biosphere is formed by inanimate bodies, which I will call inert." It was assumed that the space of living matter may be non-Euclidean. However, V.I. Vernadsky did not know what this non-Euclidean nature is, although he assumed that this property fundamentally distinguishes the space of living matter from the space of inert matter. He believed that there is an insurmountable line between living and inert matter (Vernadsky, 1977), and this line is primarily associated with the properties of the space of these substances. "Apparently, we are dealing inside organisms with a space that does not correspond to Euclid's space but corresponds to one of the forms of Riemann's space."

Only now did the intuitive guess of V.I. Vernadsky about the non-Euclidean space of living matter receive a mathematical proof (Zhizhin, 2017, 2019, 2021). Studying the geometry of various biomolecules, it has been proven that they have the highest dimension. This is the difference between the geometry of biomolecules and the geometry of Euclid. From the point of view of three-dimensional space, one can see that different figures simultaneously enter the polytopes corresponding to biomolecules, as if passing through each other. At the same time, these works convincingly proved that the molecules of inert substance also have the highest dimension. In this respect, there is no fundamental difference between living and inert matter. In the teachings of V.I. Vernadsky on the biosphere, much attention is paid to the concept of dissymmetry, introduced by P. Curie (Curie, 1908, 1966) when considering the experiments of L. Pasteur (Pasteur, 1922, 1960) on the rotation of the plane of polarization of light in solutions containing living matter. However, it has recently been shown (Zhizhin, 2019) that this property is a consequence of the higher dimensionality of the molecule of the natural chemical compound of tartaric acid, which does not belong to biomolecules. Therefore, there is no reason to consider dissymmetry as an integral property of living matter. The cornerstone of the doctrine of the biosphere is the statement about the insurmountable border between living and inert matter. Moreover, V.I. Vernadsky believed that living matter is, as a result of this insurmountable edge, under the limiting pressure from the inert matter. However, recent studies have shown (Qian, et al., 2010; Russel, et al., 2005; Simakov, et al., 2018, 2020) that as a result of the interaction of the inert substance of pyrite FS_2 with water, protocells LUCA may appear in water layers on the surface of minerals, including the surface of nano diamonds, which, as found, can form on Earth at average temperatures and pressures. It is currently assumed that all biomolecules were formed from protocells about 4.5 billion years ago. Thus, the possibility of the birth of living matter from inert matter is revealed. Moreover, it should be borne in mind that the structure of minerals, including nano diamonds, is characterized by the highest dimension (Zhizhin, 2014), as well as biomolecules (Zhizhin, 2016, 2019). In addition, water has the ability to adjust its structure to the structure of substances in its vicinity. Therefore, in the vicinity of the minerals water acquires the highest dimension (Zhizhin, 2020).

Since living and inert matter has, as a rule, the highest dimension, it is necessary to investigate the geometry of the spaces of these substances, i.e., to determine how this geometry differs from the geometry of Euclid and Riemannian geometry in its established representation. This will allow in the future to indicate in detail the features of the processes in these substances. Therefore, the monograph briefly describes the properties of Euclidean geometry and Riemannian geometry. After that, the properties of polytubes of the highest dimension of living and inert matter are described. Due to the importance of the question of the origin of life on Earth, the possibility of the origin of molecules of living matter from inert matter in protocells LUCA is considered in detail. In addition, the geometry of the interaction of nucleic acids is investigated. It is defined as the non-Euclidean nature of the polytope of hereditary information, formed in the field of interaction of nucleic acids, affecting the transmission of hereditary information. The significance of the genetic code in connection with the established laws of transmission of hereditary information in the polytope of hereditary information and their sequence in the chain of nucleotides is discussed.

Information processes in living matter play a significant role in ensuring the sustainable existence of living organisms. In this regard, the monograph pays great attention to the study of information flows in polytopes of the highest dimension, which are biomolecules. It is shown that the higher the dimensionality of the polytope, the more powerful the information flow it has. This allows living organisms to create reliable protection against harmful external influences (for example, from viruses).

In general, the monograph represents a new worldview of nature and life on Earth, which is based on the highest dimension of the molecules of chemical compounds. We can say that this is the worldview of the future, which must be studied and kept in mind in future life on Earth.

It should be noted that representatives of Russian science, to a certain extent, falsify the teachings of V.I. Vernadsky (Yanshina, 2011; Sokolov, 2013), claiming that for the first time the work of V.I. Vernadsky "Biosphere" was published in Leningrad in 1926, but this is not so. This monograph was published in 1926 by the Springer Publishing House. In addition, their descriptions of the doctrine of the biosphere do not mention the assumptions of V.I. Vernadsky about the non-Euclidean geometry in the realm of the living and the possibility of using Riemannian geometry or related geometries here. Only ideas about an insurmountable difference between living and non-living are emphasized, which at present seem insufficiently convincing.

Chapter 1

Euclidean Geometry and Non-Euclidean Geometry

Abstract

The main principles of Euclidean geometries and non-Euclidean hyperbolic and elliptic geometries are described. It is shown that the geometry of higher-dimensional polytopes, due to the geometric structures inherent in higher-dimensional polytopes, cannot be described either in the framework of the *n*-dimensional Euclidean geometry, or in the framework of non-Euclidean hyperbolic and elliptic geometries. Thus, the geometry of polytopes of the highest dimension should be regarded as a non-Euclidean geometry of a special type, not related to the curvature of space.

Keywords: higher-dimensional space, Euclidean geometry, non-Euclidean geometry, hyperbolic geometry, elliptic geometry, polytope

Introduction

The emergence of geometric representations dates back to very distant times. Their initial design is usually associated with the ancient cultures of Babylon and Egypt. From the seventh century BC, the period of the development of geometry began with the works of Greek scientists. This was especially facilitated by the philosophical schools of Plato and Aristotle. Euclid, who lived from about 330 to 275 BC, was one of the greatest geometers of antiquity. He managed to combine many facts into one whole, find proofs of various theorems and arrange them in a logical chain in the "Beginnings" he compiled. Subsequently, geometry developed significantly due to the construction of various geometric systems, the complication of the mathematical apparatus. Often these constructions were of a formal abstract nature (McMullen, Schulte, 2002; Diudea, Nagy, 2007; Ashrafi, Cataldo, Iraumanesh, Ori, 2013; Diudea, 2018; Fomenko, 1992; Bukhstaber, Panov,

2004). Within the framework of this monograph, we will pay attention to geometric systems that reflect, first of all, the real nature on planet Earth, which is based on real chemical compounds. Therefore, disregarding numerous works on the construction of abstract geometries, we will briefly describe the geometry of Euclid and the main non-Euclidean geometries.

1.1. Euclidean Geometry

Euclid's geometry, which became the basis for the study of mathematics for two millennia, is described in detail in the book by M.E. Vaschenko-Zakharchenko (1880). Its presentation can be found in the modern book Euclid. Beginnings. (2012). The basis of Euclid's geometry is five postulates (axioms that do not require proof):

1. From any point to another point, you can draw a straight line.
2. The final straight line can be continued continuously in a straight line.
3. From any center and any radius it is possible to describe a circle.
4. All right angles are equal to each other.
5. If a straight line meets two other straight lines in such a way that on one-sided they form two internal one-sided corners, in total less than two right angles, then these straight lines, if continued indefinitely, will intersect on the side from which the one-sided corners are less than two right angles.

From the fifth postulate it follows that if there are two parallel lines, i.e., the straight line intersecting them forms two inner one-sided angles to the sum of two straight lines, then these parallel straight lines will not intersect no matter how much they continue.

Euclidean geometry, according to which space is infinite, homogeneous, isotropic and three-dimensional, became the basis of classical physics. However, the fifth postulate of parallel straight lines has been in doubt since the 19th century. It is easy to make such a construction (Figure 1.1). We will draw two parallel lines, and also in the plane of these lines we will draw a circle C of an arbitrary radius centered at point A between these lines so that the circle does not intersect straight lines.

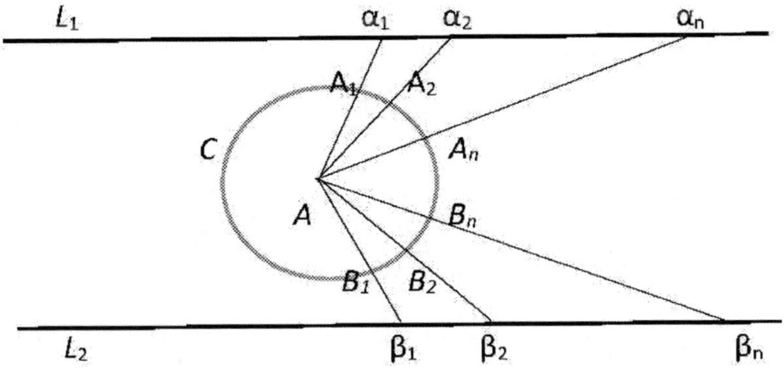

Figure 1.1. On the axiom of parallelism of straight lines.

We will draw rays from the center of the circle A until they intersect with the first (L_1) and second (L_2) lines. Let us mark the sequence of points of intersection of the rays with the line L_1 with a decrease in the angle of inclination of the rays as $\alpha_1, \alpha_2, ..., \alpha_n$

intersection of the rays with the line L_2

rays decreases as $\beta_1, \beta_2, ..., \beta_n$

intersection of the rays with the circle C as $A_1, A_2, ..., A_n$, and $B_1, B_2, ..., B_n$

we will successively decrease the angles of inclination of the rays to zero in both sets of rays. In this case, the points α_n and β_n

lines L_1 and L_2 right to infinity. The images of these points A_n B_n on the circle C will approach each other as closely as you like. At a zero angle of inclination of the rays, the point A_n will coincide with the point B_n

there is a one – to – one correspondence between the points α_n and A_n

well as between the points β_n and B_n, the alignment of the points A_n

B_n will mean the alignment of the points α_n and β_n L_1

and L_2 intersect at infinity.

The system of axioms of Euclidean geometry is generalized to n-dimensional spaces (Hilbert, Cohn-Vossen, 1932; Efimov, 1945). However, this generalization is purely mechanical. It does not take into account the structure of a higher-dimensional space, established in the works of Grünbaum (Grünbaum, 1967) and in recent years in the works of Zhizhin (Zhizhin, 2019, 2021). The inconclusiveness of the postulate of parallel lines led to the construction of geometries as a whole on the basis of concepts and axioms that relate only to the immediate vicinity of each point. So, the possibility arose of creating geometries based on differential geometry (Hilbert, Cohn-Vossen, 1932).

1.2. Non-Euclidean Hyperbolic Geometry

In non-Euclidean geometry postulate 5 of the Euclidean system of postulates is replaced by the axiom.

Through a point that does not lie on a given straight line, there are at least two straight lines that lie with this straight line in one plane and do not intersect it. This is Lobachevsky's axiom (Lobachevsky, 1829, 1835). According to this axiom, on the plane through the point P, which lies outside the straight-line $A^{/} A$, more than one straight line (Figure 1.2) does not intersect the straight-line $A^{/} A$.

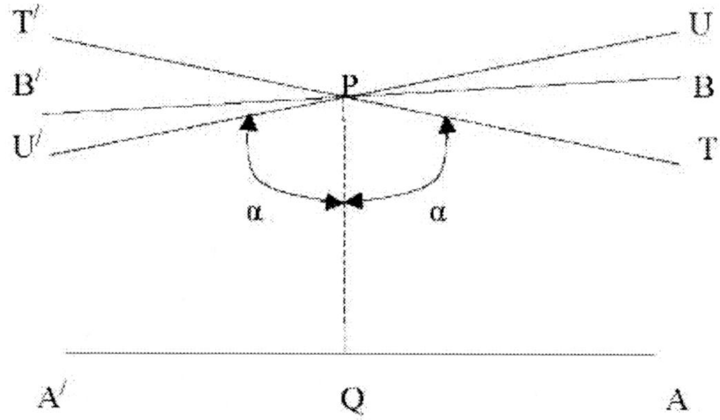

Figure 1.2. Lines passing through a point in non-Euclidean geometry.

Non-intersecting straight lines fill the part of the beam with apex P lying inside the vertical angles TPU and $U^{/}PT^{/}$ located symmetrically relative to the perpendicular PQ.

The straight lines forming the sides of the vertical corners separate the intersecting straight lines from the non-intersecting straight lines and are themselves intersecting straight lines. These boundary lines are called parallels at point P to line $A^{/} A$, respectively, in its two directions: $T^{/} T$ parallel to $A^{/} A$ in the direction of $A^{/} A$, and $U^{/} U$ parallel to $A^{/} A$ in the direction of $A^{/} A$. The rest of the non-intersecting lines are called lines diverging from $A^{/}A$. The angle α $(0<\alpha<\pi/2)$, that the parallel to the point P forms with the perpendicular PQ, $\angle QPT=\angle QPU^{/}=\alpha$ is called the angle of parallelism of the segment $PQ = a$ and is denoted by $\alpha=\Pi(a)$. When $a =$ 0 the angle $\alpha=\pi/2$; with increasing a angle α decreases so that for each given α there is a certain value of a. This dependence is called the Lobachevsky function

$$\Pi(a)=2arctg(e^{-a/k}), \tag{1.1}$$

where k is some constant that determines a segment fixed in magnitude. It is called the radius of curvature of the Lobachevsky space. There is an infinite number of Lobachevsky spaces, differing in the size of k.

The model of the Lobachevsky plane can be a surface with constant negative curvature. The simplest example of such a surface would be the pseudosphere (Figure 1.3) (Beltrami, 1868).

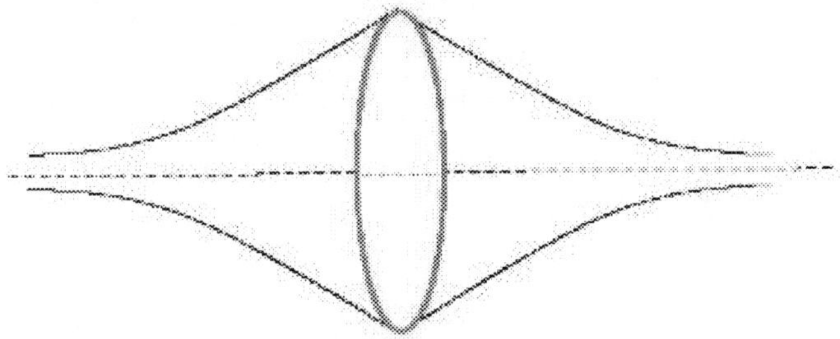

Figure 1.3. Pseudosphere.

There are also other models of Lobachevsky geometry (Klein, 1928).

1.3. Non-Euclidean Eleptical Geometry

The fact that the sphere is the spatial form of elliptic geometry is seen directly, since the curvature of a sphere of radius R in all such cases is equal to $1/R^2$, that is, it is positive and constant. It can be shown that a metric of constant curvature $K > 0$ can also be introduced on the projective plane. Consider some plane α of Euclidean space and some sphere C. Let the plane of Euclidean space be located horizontally, and the sphere is above the plane. Let O be the center of the sphere. Obviously, each point of the sphere, if it does not lie on the equator, is projected from the center O to a certain point of the plane α. Diametrically opposite points of the sphere are projected at the same point. The points lying on the equator have no projections. Supplement the plane α with infinitely distant points and introduce a system of neighborhoods into the set of all points of the extended plane. Then we get a projective plane, which we will denote by $\bar{\alpha}$. Let us agree to say that the horizontal line a passing through the point O intersects the projective plane $\bar{\alpha}$ at an infinitely distant point lying on any of the straight lines in the Euclidean plane α parallel to the line a. Under this condition, each point of the sphere is projected from the center O to a certain point of the projective plane $\bar{\alpha}$, and each point on $\bar{\alpha}$ is the projection of two diametrically opposite points of the sphere. Let us now introduce geographic coordinates on sphere C. On the projective plane, the coordinates of some point M will be considered numbers equal to the geographical coordinates of that point of the sphere, which has its projection point M. Since each point of the projective plane is the projection of two opposite points of the sphere, then each point of the projective plane, along with coordinates, also has coordinates. Now, in each neighborhood of a point on the projective plane, we introduce the metric form

$$dS^2 = R^2(\cos^2\psi d\varphi^2 + d\psi). \tag{1.2}$$

In this case, the matrix shape of the sphere is written in the form (Efimov, 1945).

$$dS^2 = \cos^2(K^{1/2}v)d^2u + d^2v \tag{1.3}$$

To reduce expression (1.2) to expression (1.3), it is sufficient to set $\varphi = K^{1/2}u, \psi = K^{1/2}v$ and take into account the relationship between the curvature and the radius of the sphere $K = 1/R^2$.

The geometric system of the metrized projective plane $\bar{\alpha}$ is called Riemannian non-Euclidean geometry. In the geometric system of the metrized projective plane $\bar{\alpha}$ (elliptic plane), any two points define one and only one geodesic line passing through them. Every two lines have a common point. An elliptical plane has no parallel lines. In an elliptical plane (Riemann geometry), every line intersects another.

The well-known objects of two-dimensional differential geometry can be generalized into multidimensional geometry (Efimov, 1945; Cartan, 2009). Riemann introduced the concept of an n-extended quantity and defined a measure definition that is possible on an n-dimensional manifold $dS^2 = \sum g_{ik} dx_i dx_k$, where the coefficients in front of the product of the differentials of the variables are functions of coordinates.

1.4. Non-Euclidean Geometry of the Higher Dimension Polytopes

Riemann pointed out the contours of geometric figures as one of the examples of the n-extended value. The space of these figures, therefore, has dimension n and is limited, i.e., certainly. The condition for space to be finite is one of the basic conditions in Riemann geometry. However, the curvature of space will be zero if the planar (two-dimensional) elements of the figure have zero curvature. Thus, there is no need to use the apparatus of differential geometry. Due to the finiteness of the higher-dimensional polytope and the presence of the polytope contour (i.e., the boundary), the higher-dimensional polytope space is closed. Therefore, the system of axioms of n-dimensional Euclidean space, extended from the system of axioms of three-dimensional Euclidean space, cannot be applied to the polytope space of higher dimension. In particular, line segments located inside the higher-dimensional polytope cannot be extended outside the higher-dimensional polytope. The space of a polytope of the highest dimension has a special structure that determines the conditions for the belonging of elements of different dimensions to each other, as will be shown in the next chapter. Therefore, the conditions for membership of elements of different dimensions, adopted in the system of axioms of n-dimensional Euclidean geometry, are not fulfilled in the polytope of the highest dimension.

In this regard, it should be recognized that the geometry of the space of polytopes of the highest dimension is not Euclidean.

Conclusion

Comparison of geometric systems of Euclidean space, as well as non-Euclidean hyperbolic and elliptic geometries is carried out. Taking into account that, according to Riemann, the space of non-Euclidean geometries is closed and, of course, the features of the geometries of polytopes of higher dimension are considered. It is shown that geometric systems of n-dimensional geometry of Euclidean cannot be used to describe polytopes of higher dimension. It follows that the geometry of higher-dimensional polytopes is non-Euclidean due to the spinal geometric structure of higher-dimensional polytopes. This non-Euclidean geometry is not related to the curvature of space and therefore cannot be attributed to either hyperbolic or elliptical non -Euclidean geometries.

Chapter 2

Geometry of the Polytopes of Higher Dimensions

Abstract

Geometric constructions of polytopes of higher dimensions of various types are considered: well-known polytopes of the n-cube, n-simplex, n-cross-polytopes type, as well as polytopes of a new type such as polytopic prisms, polytopic prismahedrons, polyincidental polytopes and polytopes dual to them. In this variety of different types of polytopes of the highest dimension, the question of whether geometric elements of a certain dimension belong to geometric elements of another dimension was investigated. The laws of incidence between geometric elements of various dimensions of the listed types of polytopes of the highest dimension are established. It was found that an increase in the dimension of the polytope can lead to a violation of its homogeneity, i.e., to the appearance of elements of a certain dimension with different geometric characteristics.

Keywords: higher-dimensional space, Euclidean geometry, non-Euclidean geometry, polytope, polytopic prismahedron, incidence

Introduction

In the first chapter it was argued that the geometry of higher-dimensional polytopes does not allow the use of the axioms of the n-dimensional geometry of Euclid, and therefore the geometry of higher-dimensional polytopes is non-Euclidean. Moreover, the geometry of higher-dimensional polytopes cannot be attributed to either hyperbolic or elliptic non-Euclidean geometries, since the non-Euclidean geometry of higher-dimensional polytopes is not associated with the curvature of space, as is the case in hyperbolic and elliptical geometries, but is determined by the peculiarities of the geometric structure of higher-dimensional polytopes. In this regard, this chapter gives a detailed presentation of the features of the geometric structure of polytopes of the highest dimension of different types, which was absent in the first chapter since the main goal of the first chapter was to recall the essence of Euclidean geometry and non-Euclidean geometries associated with the curvature of space. The structure of n-dimensional polytopes of the cube, simplex and cross-polytope types is considered. The conditions for the existence of internal points of these polytopes are established, and the structure of their boundary complexes is determined. Laws are formulated that determine the quantitative presence of geometric elements of various dimensions in the boundary complexes. The laws of incidence in these structures are also formulated, which determine the order of incidence of geometric elements of a certain dimension to geometric elements with a dimension greater by one. The sequence of constructing n-cube, n-simplex, n-cross-polytopes according to the established laws for any dimension of polytopes is demonstrated. The structure of a cube, octahedron and tetrahedron is investigated with the addition of vertices to their centers. The quantitative characteristics of the structures of polytopic prismahedrons, which are the product of polytopes, are investigated. The structure of polytopes dual to polytopic prismahedrons is investigated. It is shown how the construction of a dual polytope to a given polytopic prismahedron leads to a violation of the homogeneity of the structure, i.e., to the appearance of geometric elements of a certain dimension with different incidence values. It has been found that an increase in the dimension of a polytope can itself lead to a violation of the homogeneity of the structure (the appearance of poly-incident polytopes).

In the third chapter, a detailed comparison of the geometric features of polytopes of higher dimension with the axioms of the n-dimensional geometry of Euclid will be carried out.

2.1. Structure of the Polytope Higher Dimension

In all geometric systems, when listing the axioms, conditions are indicated for the belonging of some elements of a lower dimension to geometric elements of a higher dimension. For example, two points belonging to a straight line define it; three points of the plane define it, etc. If, following Riemann (Riemann, 1854), we consider a geometric figure to be a limited and closed space, then in a polytope of higher dimensions there are many geometric elements of various dimensions. The structure of a polytope will be determined by the conditions for belonging of geometric elements of a lower dimension to geometric elements of a higher dimension. Many of these elements form a boundary complex.

A finite family C of convex polytopes will be called a complex under the condition

1. each face of an element of the family C is an element of the family C,
2. the intersection of any two elements of C is a face of both.

Let P is an n-polytope, i.e., a convex polytope of dimension n. We denote by B (P) the boundary complex P, that is, a complex consisting of all faces of P with dimension n-1 or less (Grunbaum, 1967). The points not belonging to B (P) will be called interior points of P. The set of interior points of the polytope P will be denoted by $V(P)$. It should be emphasized that the elements of the boundary complex are polytypes of dimension less than n, i.e., they are also polytopes. They also have boundary complexes and interior points. Let F_ξ P, the dimension of the face is equal to i_F

$B(F)$. The interior points of the face F_ξ will be denoted by $V(F)$ respectively the interior points of the polytope $P$are $V(P)$

the interior points of the polytope P and the interior points of the face F_ξ have no common elements. However, the faces of the polytope are involved in the interaction of polytopes with each other. In this case, the internal points of the face act as part of the face and are, together with the rest of the face, common elements of two interacting polytopes. In this case, the inner points of the face are not the inner points of the polytope. Thus, we have equality $P = B(P) \bigcup V(P)$

dimension is determined by the ratio of faces of different dimensions to each other in the boundary complex B (P).

For each element b_j of the boundary complex B (P) can introduce of the notation its dimension $i_{b_j}, 0 \le i_{b_j} < n.$

There is denoted k_{b_j}

complex b_j to the elements of the boundary complex having dimension $i_{b_j}+1$. The value k_{b_j}

with the dimension $i_{b_j}+1$ the element b_j with the dimension i_{b_j}

2.1.1. The Structure of a N-Cube

In the n-cube, the number of elements of dimension i (i = 0, 1, 2, ..., n) is determined by the formula (Zhizhin, 2019),

$$f_i(n) = 2^{n-i} C_n^i, i = 0,1,2,...,n. \tag{2.1}$$

and the coefficient of incidence of elements b_j with dimension i is determined by dependence

$$k_{b_j}(n) = n - i. \tag{2.2}$$

In accordance with these dependencies, the values $k_{b_j}(n)$, $f_i(n)$ functions of i are presented in Table 2.1.

From Table 2.1 it follows that the incidence coefficient of elements in an n-cube with respect to elements with a dimension of one more linearly decreases from n for vertices to 1 for facets. The number of elements with dimension i with growth i first grows and then falls. With an increase in the dimension of the n-cube, both quantities $k_{b_j}(n)$ and $f_i(n)$ $k_{b_j}(n)$ grows linearly, and $f_i(n)$

Table 2.1. The functions $k_{b_j}(n)$, $f_i(n)$ i for n-cube

i	$k_{b_j}(n)$	$f_i(n)$
0	n	2^n
1	n-1	$2^{n-1}C_n^1$
2	n-2	$2^{n-2}C_n^2$
3	n-3	$2^{n-3}C_n^3$
.	.	.
.	.	.
.	.	.
n-1	1	2n

2.1.2. The Structure of a N-Simplex

In the n-simplex, the number of elements of dimension i (i = 0, 1, 2, ..., n) is determined by the formula (Zhizhin, 2019),

$$f_i(n) = C_{n+1}^{i+1}, i = 0, 1, 2, ..., n. \tag{2.3}$$

and the coefficient of incidence of elements b_j with dimension i is determined by dependence (2.2). In accordance with these dependencies, the values $k_{b_j}(n)$, $f_i(n)$ i are presented in Table 2.2.

Table 2.2. The functions $k_{b_j}(n)$, $f_i(n)$ i for n-simplex

i	$k_{b_j}(n)$	$f_i(n)$
0	n	$n+1$
1	n-1	C_{n+1}^2
2	n-2	C_{n+1}^3
3	n-3	C_{n+1}^4
.	.	.
.	.	.
.	.	.
n-1	1	n + 1

From Table 2.2 it follows that the incidence coefficient of elements in an n-simplex as well as in an n-cube with respect to elements with a dimension of one more linearly decreases from n for vertices to 1 for facets. The number of elements with dimension i with growth i first grows and then falls binomial. With an increase in the dimension of the n-simplex, both quantities $k_{b_j}(n)$ and $f_i(n)$

2.1.3. The Structure of a N-Cross-Polytope

In the n-cross-polytope the number of elements of dimension i (i = 0, 1, 2, ..., n) is determined by the formula (Zhizin, 2019)

$$f_i(n) = 2^{1+i} C_n^{n-1-i}, i = 0,1,2,...,n, \tag{2.4}$$

and the coefficient of incidence of elements b_j with dimension i is determined by dependences

$$k_{b_j}(n) = 2(n-1-i), i < n-1,$$
$$k_{b_j}(n) = 1, i = n-1. \tag{2.5}$$

In accordance with dependencies (2.4), (2.5), the values $k_{b_j}(n)$, $f_i(n)$ as functions of i are presented in Table 2.3.

From Table 2.3 it follows that the incidence coefficient of elements in an n-cross-polytope with respect to elements with a dimension of one more linearly decreases from 2(*n-1)* for vertices to 1 for facets.

Table 2.3. The functions $k_{b_j}(n)$, $f_i(n)$ i for n-cross-polytope

i	$k_{b_j}(n)$	$f_i(n)$
0	2(n-1)	$2n$
1	2(n-2)	$2^2 C_n^{n-2}$
2	2(n-3)	$2^3 C_n^{n-3}$
3	2(n-4)	$2^{n-4} C_n^{n-4}$
.	.	.
.	.	.
.	.	.
n-1	1	2^n

For all values of dimension i besides n-1 the incidence coefficient is significantly larger than that of an n-cube and n-simplex. The number of elements with dimension i is larger than of an n-cube and n-simplex too.

2.1.4. Reconstruction of the Geometry of Polytopes of Higher Dimension

According to Tables 2.1-2.3 and equations (2.1)-(2.5), one can easily determine the characteristics of polytopes of the highest dimension, depending on their dimension. As a beginning, let the dimension of the polytopes be 3. Then the polytope of the cube type (see Table 2.1) has 8 vertices. There are 3 edges emanating from each vertex, each edge is common to two flat faces, each flat face belongs to one 3-cube. A three-dimensional cube has 12 edges, 6 flat faces. When the dimension is 3, then the 3-simplex (see Table 2.2) has 4 vertices. 3 edges emanate from each vertex, each edge is common for 2 flat faces, each flat face belongs to one 3-simplex, 3-simplex has 6 edges, 4 flat faces. This figure is a tetrahedron.

When the dimension is 3, then the 3-cross-polytope (see Table 2.3) has 6 vertices, 12 edges, 4 edges emanate from each vertex, each edge is common to two flat faces, each flat face belongs to one 3-cross-polytope. This polytope has 8 flat faces. This figure is an octahedron.

Similar calculations can be done for any dimension.

Let the dimension of the considered polytopes be 6. Then the polytope 6-the cube has (see Table 2.1) 64 vertices, 6 edges emanate from each vertex, the total number of edges is 192, each edge is a common 5 flat faces, the total number of flat faces is 240. Each flat face is a common four three-dimensional faces (3-cubes). The total number of three-dimensional faces is 160, each three-dimensional face is a common three four-dimensional faces (4-cubes). The total number of 4-cubes is 60, each 4-cube is common to two 5-cubes (5-cubes). The total number of 5-cubes is 12.

When the dimension is 6, then the 6-simplex has (see Table 2.2) 7 vertices, 6 edges emanate from each vertex, the total number of edges is 21. Each edge is common to five flat faces. The total number of flat faces is 35. Each flat face is common to four three-dimensional faces (tetrahedrons). The total number of tetrahedra is 35. Each tetrahedron is common to three 4 -simplex polytopes. The total number of 4-simplex polytopes is 21. Each 4-simplex polytope is common to two 5-simplex polytopes. The total number of 5-simplex polytopes is 7.

In the case of an *n*-cross-polytope, it should be remembered that an *n*-cross-polytope is a simplicial polytope, and all its faces are simplexes. When the cross-polytope has dimension 4 (4-cross-polytope), then the 4-cross-polytope (see Table 2.3) has 8 vertices, 6 edges emanate from each vertex, the total number of edges is 24, each edge is a common four two-dimensional flat faces, the total number of two-dimensional faces is 32, each flat face is common to two three-dimensional faces (tetrahedrons), the total number of three-dimensional faces is 16.

When a cross-polytope has dimension 6 (6-cross-polytope), then it has (see Table 2.3) 12 vertices, from each vertex there are 10 edges. The total number of edges is 60. Each edge is common to 8 flat faces. The total number of flat faces is 160. Each flat face is a common six three-dimensional faces (tetrahedrons). The total number of tetrahedra is 240. Each tetrahedron is common to four 4-simplex polytopes. The total number of 4-simplex polytopes is 192. Each 4-simplex is common to two 5-simplex polytopes. The total number of 5-simplex polytopes is 64.

In the same way, using Tables 2.1-2.3, it is possible to determine the structure of polytopes of the types a cube, simplex, and cross-polytope for any of their dimensions.

2.2. The Structure of the Cube with Center

Consider a cube with center at point a_0 (Figure 2.1).

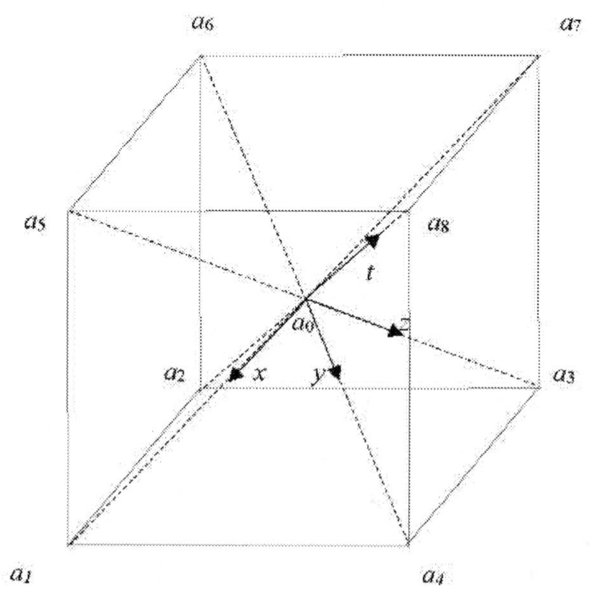

Figure 2.1. Cube with center.

It is denoted the vertices of the cube $a_j (j = 1 \div 8)$
center of the diagonals of the cube. Let's enumerate the geometric elements of different dimensions of this construction: 9 vertices $(f_0 = 9)$
$(f_1 = 20)$:

$$e_1 = a_1a_2, e_2 = a_3a_2, e_3 = a_3a_4, e_4 = a_1a_4, e_5 = a_5a_6, e_6 = a_6a_7, e_7 = a_7a_8, e_8 = a_5a_8, e_9 = a_1a_5,$$
$$e_{10} = a_4a_8, e_{11} = a_6a_2, e_{12} = a_3a_7, e_{13} = a_0a_1, e_{14} = a_0a_2, e_{15} = a_0a_3, e_{16} = a_0a_4, e_{17} = a_0a_5,$$
$$e_{18} = a_0a_6, e_{19} = a_0a_7, e_{20} = a_0a_8;$$

18 two-dimensional faces ($f_2 = 18$): 6 square faces –

$$s_1 = a_1a_2a_3a_4, s_2 = a_5a_6a_7a_8, s_3 = a_1a_5a_8a_4, s_4 = a_6a_2a_3a_7, s_5 = a_1a_2a_5a_6, s_6 = a_7a_8a_3a_4;$$

12 triangular faces –

$$s_7 = a_0a_1a_4, s_8 = a_0a_3a_4, s_9 = a_0a_2a_3, s_{10} = a_0a_1a_2, s_{11} = a_0a_5a_8, s_{12} = a_0a_5a_6, s_{13} = a_0a_6a_7,$$
$$s_{14} = a_0a_8a_7, s_{15} = a_0a_6a_2, s_{16} = a_0a_1a_5, s_{17} = a_0a_3a_7, s_{18} = a_0a_8a_4;$$

7 three-dimensional faces ($f_3 = 7$):

1 cube, 6 pyramids –

$$v_2 = a_0a_1a_2a_3a_4, v_3 = a_0a_5a_6a_7a_8, v_4 = a_0a_1a_2a_5a_6, v_5 = a_0a_7a_8a_3a_4, v_6 = a_0a_1a_5a_8a_4, v_7 = a_0a_6a_2a_3a_7.$$

We will check the existence of a polytope of higher dimension using the Euler-Poincaré equation (Poincaré, 1895)

$$\sum_{i=0}^{n-1} (-1)^i f_i(P) = 1 + (-1)^{n-1}. \tag{2.6}$$

Substituting the obtained quantities of elements of different dimensions into the Euler-Poincaré equation (2.6), can obtain

9 - 20 + 18 - 7 = 0.

This means that equation (2.6) is satisfied for n = 4. Consequently, the cube with the center has dimension 4 (Zhizhin, Diudea, 2017).

The incidence of the elements of the boundary complex in this case is determined by the equalities

$$k_{a_0}(4)=8; k_{a_j}(4)=4, j=1\div 8; k_{e_j}(4)=3, j=1\div 20; k_{s_j}(4)=2, j=1\div 18; k_{v_j}(4)=1, j=1\div 7$$

Since the cube with the center has dimension 4, it is necessary to place the origin of coordinates of the four-dimensional system in it. It is convenient to take the center of the cube as the origin of coordinates by sending coordinates along the diagonals of the cube (Figure 2.1).

Let's take a guess from the center of the cube to any of its vertices for a unit. Then all the vertices of polytope are determined by the values of the coordinates (x, y, z, t)

$$a_0(0,0,0,0), a_1(1,0,0,0), a_2(0,0,0,-1), a_3(0,0,1,0), a_4(0,1,0,0), a_5(0,0,-1,0),$$
$$a_6(0,-1,0,0), a_7(-1,0,0,0), a_8(0,0,0,1).$$

Each of the facets of a boundary complex is determined by the corresponding system of inequalities

$$v_1: -1\le x\le 1, -1\le y\le 1, -1\le z\le 1, -1\le t\le 1; v_2: 0\le x\le 1, 0\le y\le 1, 0\le z\le 1, -1\le t\le 0;$$
$$v_3: -1\le x\le 0, -1\le y\le 0, -1\le z\le 0, 0\le t\le 1; v_4: 0\le x\le 1, -1\le y\le 0, -1\le z\le 0, -1\le t\le 0;$$
$$v_5: -1\le x\le 0, 0\le y\le 1, 0\le z\le 1, 0\le t\le 1; v_6: 0\le x\le 1, 0\le y\le 1, -1\le z\le 0, 0\le t\le 1;$$
$$v_7: -1\le x\le 0, -1\le y\le 0, 0\le z\le 1, 0\le t\le 1.$$

Thus, the entire polytope it is occupied by its facets, i.e., its boundary complex, the polytope does not have interior points. If one recall that the boundary complex is used as a synonym for the surface of the figure, then we came to an amazing result: the surface of the cube with the center completely occupies the space of this polytope.

2.3. The Structure of the Octahedron with Center

The octahedron with the center has 7 vertices (Figure 2.2) $a_0 \div a_6, (f_0 = 7)$; 18 edges $(f_1 = 18)$-

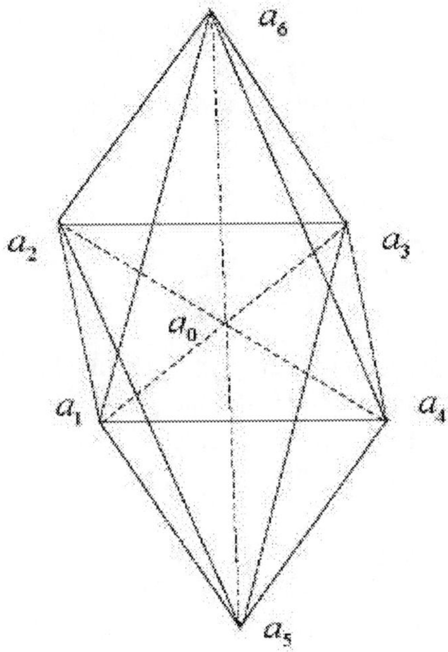

Figure 2.2. The octahedron with center.

$e_1 = a_0a_3, e_2 = a_0a_4, e_3 = a_0a_1, e_4 = a_0a_2, e_5 = a_0a_6, e_6 = a_0a_5, e_7 = a_1a_6, e_8 = a_2a_6, e_9 = a_6a_3,$
$e_{10} = a_4a_6, e_{11} = a_1a_5, e_{12} = a_2a_5, e_{13} = a_4a_5, e_{14} = a_5a_3, e_{15} = a_1a_2, e_{16} = a_2a_3, e_{17} = a_4a_3, e_{18} = a_1a_4,;$

20 flat two-dimensional trigonal faces $(f_2 = 20)$-

$s_1 = a_1a_2a_6, s_2 = a_3a_2a_6, s_3 = a_3a_4a_6, s_4 = a_1a_4a_6, s_5 = a_1a_2a_5, s_6 = a_3a_2a_5, s_7 = a_3a_4a_5,$
$s_8 = a_1a_4a_5, s_9 = a_1a_0a_6, s_{10} = a_0a_2a_6, s_{11} = a_0a_3a_6, s_{12} = a_0a_4a_6, s_{13} = a_1a_0a_5, s_{14} = a_0a_4a_5,$
$s_{15} = a_0a_3a_5, s_{16} = a_0a_2a_5, s_{17} = a_1a_0a_4, s_{18} = a_1a_2a_0, s_{19} = a_0a_2a_3, s_{20} = a_0a_3a_4;$

9 three-dimensional faces $(f_3 = 9)$: 1 the octahedron, 8 the pyramids

$v_2 = a_1a_0a_4a_5, v_3 = a_3a_0a_4a_5, v_4 = a_3a_0a_2a_5, v_5 = a_1a_0a_2a_5, v_6 = a_1a_0a_4a_6, v_7 = a_3a_0a_4a_6,$
$v_8 = a_3a_0a_2a_6, v_9 = a_1a_0a_2a_6.$

Substituting the obtained quantities of elements of different dimensions into the Euler-Poincaré equation (2.6), can obtain

7-18 + 20-9 = 0.

This means that equation (2.6) is satisfied for $n = 4$. Consequently, the octahedron with the center has dimension 4.

The incidence of the elements of the boundary complex in this case is determined by the equalities

$$k_{a_0}(4)=6; k_{a_j}(4)=5, j=1\div 6; k_{e_j}(4)=4, j=1\div 6; k_{e_j}(4)=3, j=7\div 18;$$
$$k_{s_j}(4)=2, j=1\div 20; k_{v_j}(4)=1, j=1\div 9.$$

Now can see, without introducing the coordinate system (Figure 2.2), that an octahedron with a center does not have interior points, since the entire neighborhood of the center is filled with facets.

2.4. The Structure of the Tetrahedron with Center

The tetrahedron with the center has 5 vertices (Figure 2.3) $a_0 \div a_5, (f_0 = 5)$; 10 edges $(f_1 = 10)$, 10 two-dimensional trigonal faces $(f_2 = 10)$, and 5 three-dimensional faces $(f_3 = 5)$.

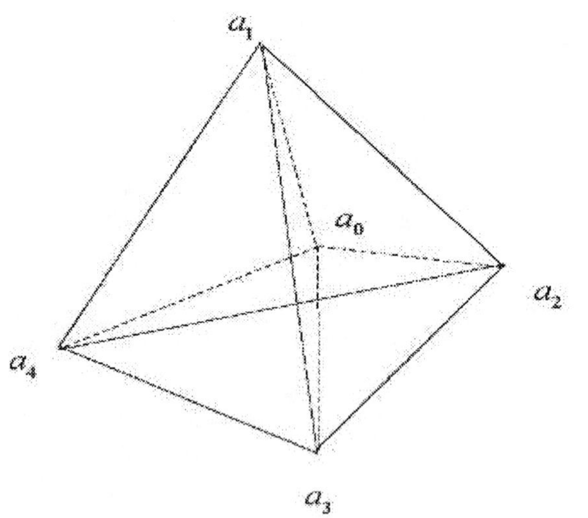

Figure 2.3. The tetrahedron with center.

The number of elements of different dimensions in this boundary complex coincide with the values of these quantities in the polytope 4-simplex. Therefore, can assume that the dimension of a tetrahedron with center is 4.

The incidence of the elements of the boundary complex in this case coincide with the values of these quantities in the polytope 4-simplex too

$$k_{a_j}(4)=4, j=0\div 5; k_{e_j}(4)=3, j=1\div 10; k_{s_j}(4)=2, j=1\div 10; k_{v_j}(4)=1, j=1\div 5.$$

Therefore, can assert that a tetrahedron with a center is topologically equivalent to a polytope 4-simplex. We have previously proved that the polytope 4-simplex has interior points. Consequently, a tetrahedron with its center has interior points. The boundary complex in it does not occupy this figure entirely. An indicator of the existence of internal points in a polytope of higher dimension can be the homogeneity of elements of different dimensions in the boundary complex. In a cube with a center and an octahedron centered, this homogeneity was broken, vertices appeared with different incidence values. This indicated that it is not impossible to deform continuously the figure to obtain a uniform distribution of vertices with the same incidence values, in which there may be interior points.

2.5. Polytopic Prismahedrons

In the monograph Zhizhin G.V. (Zhizhin, 2017) it was established that the fundamental area in the filling of the n-dimensional space is the polytopic prismahedron, which is a prism or a complex of prisms with bases in the form of polytopes of dimension n. It is polytopes provide the filling of the n-dimensional space with a face into the face without gaps (Zhizhin, 2021). Thus, the polytopic prismahedron is the stereohedron, the existence of which was suggested by B.N. Delone (Delone, 1961; Delone, Sandakova, 1961). However, Delone never succeeded in constructing any particular stereohedron in a space with a dimension greater than three. Pontryagin mentioned the structures resulting from the product of a polyhedron by a one-dimensional segment the cylinder ones (Pontryagin, 1976). We can say that the product of a polytope by a segment is a prism (Robertson, 1984) with a base in form of a polytope. To distinguish it from a usual three-dimensional prism, can call it polytopic prism. Ziegler noted that the product of polytopes is not a simplex even if the factors are simplexes, so the polytopes are of considerable interest (Ziegler, 1995). Especially taking into account that a multi-dimensional world has its own peculiarities having no analogues in the three-dimensional world (Ziegler, 1995). In this regard, the theory of simplicial polytopes for the analysis of polytopes product becomes inapplicable, especially in the case of high-dimensional factors.

The product of two polytopes is result of the product of one of them by one-dimensional edges of another polytope. Thus, the product of polytopes is a complex of polytopic prism. Let`s call this complex a polytopic prismahedron. The polytopic prism is particular case of polytopic prismahedron. The interest to the study of polytopic prismahedrons is connected, in first place, with the novelty of this field and, secondly, with the fact that polytopic prismahedrons, due to their construction, can be "bricks" to fill the spaces of higher dimension face in face.

The definition of polytopes product (Ziegler, 1995) does not give the possibility to specify the structure of product of polytopes as a function of the factor structures (Zhizhin, 2019).

2.6. Structure of Polytopic Prisms

We consider in more detail the geometric properties of polytopic prisms. Let's turn to the simplest of polytopic prisms-the tetrahedral prism (Figure 2.4) and decide whether the tetrahedral prism has internal points, i.e., points not belonging to its boundary complex.

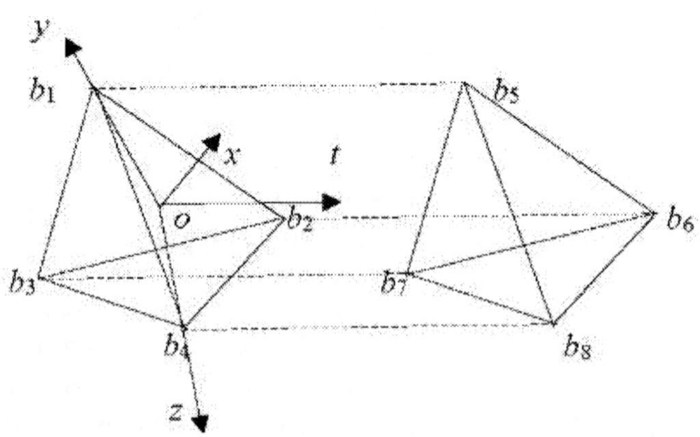

Figure 2.4. The tetrahedral prism.

For this purpose, in Figure 2.4, coordinate systems are introduced, as well as when the tetrahedron is considered. Taking into account that in this case we are dealing with a space of dimension 4, a fourth coordinate *t* is added to the coordinate system parallel to the generating lines of the tetrahedral prism. Assuming the length of all edges of the tetrahedral prism per unit, one can establish the coordinates of all the vertices of the tetrahedral prism (Figure 2.4) in the given coordinate system (*x, y, z, t*)

$$b_1\left(0,\frac{1}{\sqrt{3}},0,0\right), b_2\left(\frac{1}{2},-\frac{1}{2\sqrt{3}},0,0\right), b_3\left(-\frac{1}{2},-\frac{1}{2\sqrt{3}},0,0\right), b_4\left(0,0,\sqrt{\frac{2}{3}},0\right),$$

$$b_5\left(0,\frac{1}{\sqrt{3}},0,1\right), b_6\left(\frac{1}{2},-\frac{1}{2\sqrt{3}},0,1\right), b_7\left(-\frac{1}{2},-\frac{1}{2\sqrt{3}},0,1\right), b_8\left(0,0,\sqrt{\frac{2}{3}},1\right).$$

Then the set of interior points of the tetrahedral prism is determined by the system of inequalities

$$z>0, 0<t<1;$$

$$z<2\sqrt{\frac{2}{3}}x-\sqrt{2}y+\sqrt{\frac{2}{3}}, -\frac{1}{2}<x<0, -\frac{1}{2\sqrt{3}}<y<\frac{1}{\sqrt{3}}, 0<\mathrm{t}<1;$$

$$z<2\sqrt{2}y+\sqrt{\frac{2}{3}}, -\frac{1}{2}<x<\frac{1}{2}; -\frac{1}{2\sqrt{3}}<y<0, 0<t<1;$$

$$z<-\sqrt{6}x-\sqrt{2}y+\sqrt{\frac{2}{3}}, 0<x<\frac{1}{2}, -\frac{1}{2\sqrt{3}}<y<\frac{1}{\sqrt{3}}, 0<t<1.$$

Similarly, in principle, one can define a set of internal points and other polytopic prisms.

We find the incidence of the elements of a boundary complex of different dimensions of a tetrahedral prism

$$k_{b(i=0)}(4)=4, k_{b(i=1)}(4)=3, k_{b(i=2)}(4)=2, k_{b(i=3)}(4)=1.$$

Thus, the incidence values of the elements of the boundary complex of the tetrahedral prism obey the formula (2.2).

The same incidence values have elements of the boundary complex of the pyramidal prism (Figure 2.5) with one exception: the tops of the Egyptian pyramids have an incidence value

$$k_{b(i=0)}(4) = 5.$$

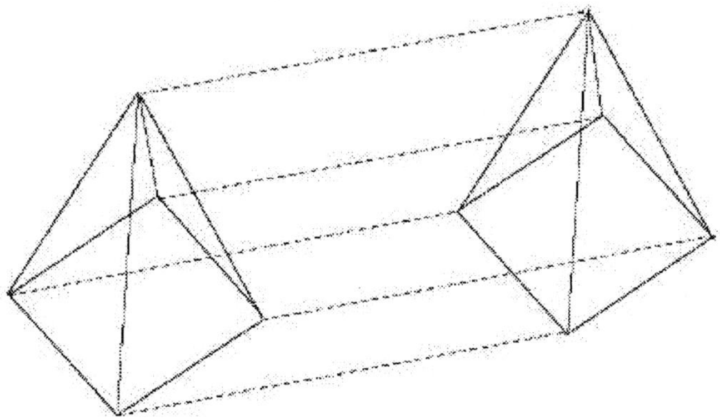

Figure 2.5. The pyramidal prism.

The polytopic prisms with increasing dimension, if they do not include cross-polytopes, also formula (2.2). For example, the 5-simplex-prism (Figure 2.6) has the following incidence of elements of its boundary complex

$$k_{b(i=0)}(5) = 5, k_{b(i=1)}(5) = 4, k_{b(i=2)}(5) = 3, k_{b(i=3)}(5) = 2, k_{b(i=4)}(5) = 1.$$

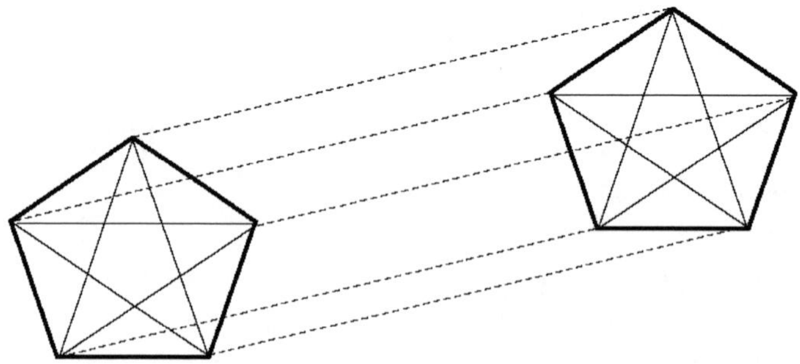

Figure 2.6. The 5-simplex-prism.

This agrees with formula (2.2).

However, if the cross-polytope is included in the boundary polytope prism complex, the situation changes. For example, Figure 2.7 shows an image of a 5-cross-prism, which includes two the 4-cross-polytopes.

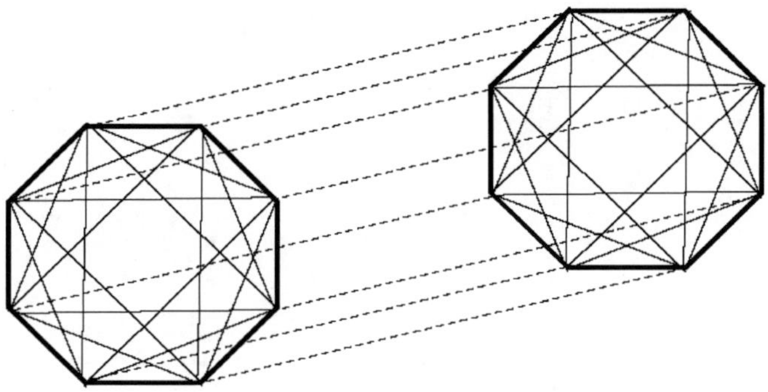

Figure 2.7. The 5-cross-prism.

From this image it follows that in this case the incidence of the elements of the boundary complex has the values

$$k_{b(i=0)}(5)=7, k_{b(i=1)}(5)=\begin{cases}6\\5\end{cases}, k_{b(i=2)}(5)=4, k_{b(i=3)}(5)=2, k_{b(i=4)}(5)=1.$$

The maximum value of incidence of edge 6 refers to the forming lines of the 5-cross-prism, the minimum incidence value of edge 4 refers to the edges of the 4-cross-polytopes.

2.7. The Incidence Values in the Polytopic Prismahedrons

The incidence of elements of the boundary complexes of polytopic prisms is also preserved for polytopic prismahedrons. For many of them formula (2.2) is satisfied. For example, for a polytopic prismahedron that is a product of a square and a tetrahedron (Figure 2.8) the incidence values of elements of a boundary complex of different dimensions have the form

$$k_{b(i=0)}(5)=5, k_{b(i=1)}(5)=4, k_{b(i=2)}(5)=3, k_{b(i=3)}(5)=2, k_{b(i=4)}(5)=1.$$

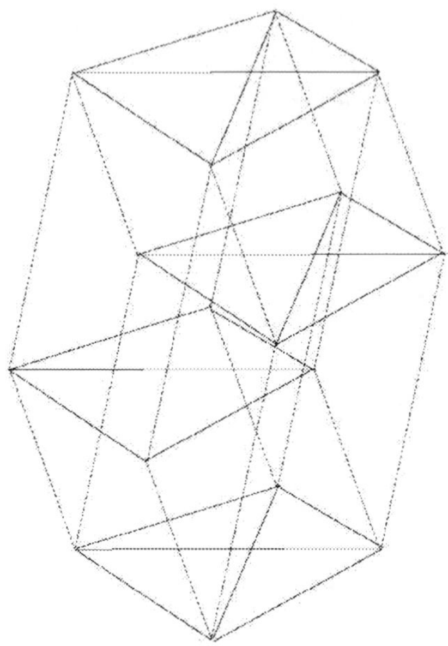

Figure 2.8. The product of tetrahedron by a square.

This completely coincides with formula (2.2).

For the polytopic prismahedron, which is the product of a tetrahedron on a tetrahedron (Figure 2.9), the incidence values of elements of a boundary complex of different dimensions have the form

$$k_{b(i=0)}(6)=6, k_{b(i=1)}(6)=5, k_{b(i=2)}(6)=4, k_{b(i=3)}(6)=3, k_{b(i=4)}(6)=2, k_{b(i=5)}(6)=1.$$

This also agrees with formula (2.2).

However, for polytopic prismahedron in which a pyramid participates (or an octahedron that can be considered as a double pyramid), there is a difference from the formula (2.2). This difference was already observed when considering the product of the Egyptian pyramid on a one-dimensional segment (Figure 2.5). The incidence of vertices in this case is greater than the value determined by formula (2.2) For example, for a polytopic prismahedron that is the product of an octahedron per triangle (Figure 2.10), the incidence values of the elements of the boundary complex have the form

$$k_{b(i=0)}(5)=6, k_{b(i=1)}(5)=4, k_{b(i=2)}(5)=3, k_{b(i=3)}(5)=2, k_{b(i=4)}(5)=1.$$

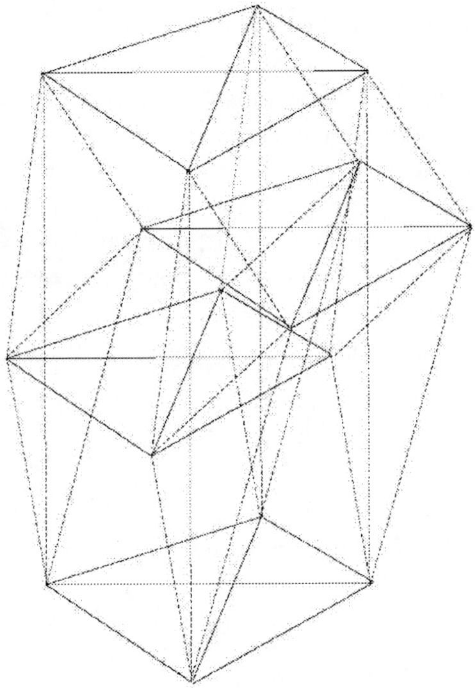

Figure 2.9. The product of a tetrahedron by a tetrahedron.

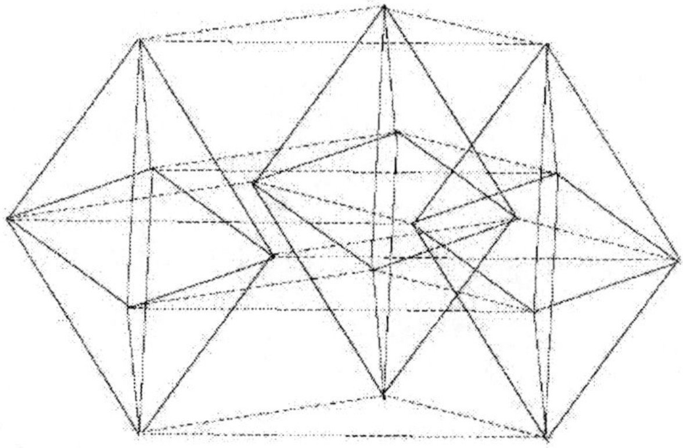

Figure 2.10. The octahedral prismahedron.

The incidence of vertices in this case is greater than the dimension of the polytopic prismahedron.

2.8. Poly-Incident and Dual Polytopes

It was convinced that the coefficient of incidence of the vertices of the Egyptian pyramid and polytopic prismahedrons with the participation of pyramids in their boundary complexes can have simultaneously several different values in the same polytope. It turns out that polytopic prismahedrons lead to polytopes with different incidence values not only of vertices, but also of other elements of boundary complexes and in the absence of pyramids.

Let's denote $b(i)$ -an element of dimension *i*; $k_{b(i),b(j)}$ -the value of the incidence of the element with dimension *i* in relation to the elements of dimension *j* (*j* > *i*). For regular polytopes because of their uniform values of the incidence $k_{b(i),b(j)}$ are constant for the whole polytope in all dimensional range from 0 to *n* (*n*-dimension of the polytope). Obviously, that $k_{b(i),b(n)} = 1$ for any $i < n$.

In a polygon there are $b(0) \in 2b(1); b(1) \in b(2)$.

In a polyhedron there are
$b(3) : b(0) \in k_{b(0),b(1)} b(1); k_{b(0),b(1)} = 3, 4, 5; b(1) \in 2b(2); b(2) \in b(3)$.

In four-dimensional polytopes relations of the incidence have the following values.

In a 4-simplex:

$$b(0) \in 4b(1); b(1) \in 3b(2); b(2) \in 2b(3); b(3) \in b(4); b(0) \in 4b(3); b(0) \in 4b(2).$$

In a 4-cube:

$$b(0) \in 4b(1); b(1) \in 3b(2); b(2) \in 2b(3); b(3) \in b(4); b(0) \in 3b(3); b(0) \in 6b(2); b(1) \in 3b(3).$$

In a 4-cross-polytope:

$$b(0) \in 6b(1); b(1) \in 4b(2); b(2) \in 2b(3); b(3) \in b(4); b(0) \in 8b(3); b(0) \in 10b(2); b(1) \in 4b(3).$$

In the construction of dual polytopes to polytopes of dimension *n*, the incidence coefficients with respect to facets of polytopes $k_{b(i),b(n-1)}$. For correct four-dimensional polytopes, these incidence coefficients have, respectively, the values:

In a 4-simplex $k_{b(0),b(3)} = 4; k_{b(1),b(3)} = 3; k_{b(2)b(3)} = 2;$

In a 4-cube $k_{b(0),b(3)} = 4; k_{b(1),b(3)} = 3; k_{b(2)b(3)} = 2;$

In a 4-cross-polytope $k_{b(0),b(3)} = 8; k_{b(1),b(3)} = 4; k_{b(2)b(3)} = 2.$

In semi-regular polytopes, relations of incidence keep their form the same as in regular polytopes (Zhizhin, 2014). But there different figures in one polytope may serve as elements *b*(2), though all vertices of these semi-regular polytopes are superposed by motion. If in a polytope there are vertices which are not superposed by motion, then relations of incidence are variable in a polytope, for example, in a triangle prism. A prism can be considered a semi-regular polytope because it has two triangle faces and three of square faces.

For a deeper study of the geometry of polytopic prismahedron, we apply the analysis of the duality of polytopic prismehedron. The analysis of duality, as an analysis of the unity of opposites, is used in various fields of science: in mathematics, logic, philosophy, physics and chemistry (Bogdanov, 1989; Whitehead, 1990; Feynman, 1968; Hegel; 1998; Bucur and Delanu, 1972).

One will consider the polytope as dual to a given polytope of dimension *n* if the facets of a given polytope are replaced on the vertices of the polytope. Moreover, the vertices of a new polytope are connected by an edge if the corresponding facets have a common face of dimension *n*-2 in the given polytope. Consider a rectangular prism with triangular bases. Can introduce the centers of the planar faces of this prism and join the edges of those centers whose faces have a common edge. One will get a double pyramid (Figure 2.11). It is an irregular polyhedron dual to this prism.

In a double pyramid two vertices are incident to three edges, and another two vertices are incident to four edges. The same picture will be if we take a pentagonal prism.

Two vertices of double pentagonal pyramid are incident to 5 edges, and 5 of the remaining vertices are incident to 4 edges (Figure 2.12).

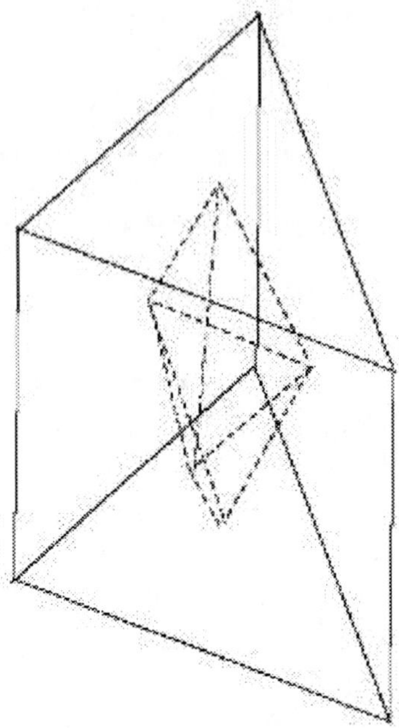

Figure 2.11. The triangle prism and double pyramid dual to it.

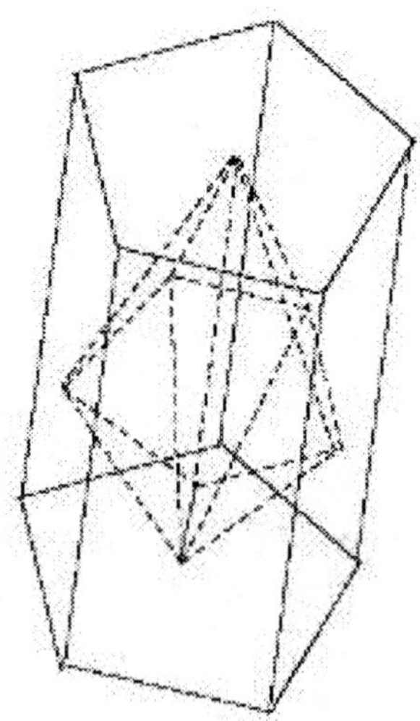

Figure 2.12. The pentagonal prism and a double pyramid dual to it.

Further it will be shown that the polytopes dual to polytopes products have not only vertices with different values of incidence to the edges, but the edges with different values of incidence to three-dimensional figures. This new type of polytopes is called poly-incident polytopes.

2.8.1. Polytope Dual to the Product of Two Triangles

In book Zhizhin (2019) the structural formula of the product of two triangles was obtained, according to which this 4-polytope has 9 vertices and 6 triangular prisms act as facets (Figure 2.13). This polytope is a triangular prismahedron. It has 18 edges, 9 squares and 6 triangles as two-dimensional elements. For clarity, one type of triangles of the two types as factors in Figure 2.13 are indicated by solid lines, and the other type of triangles are indicated by dashed lines. Let's denote vertices of the polytope by symbols $a_1, \ldots, a_9$

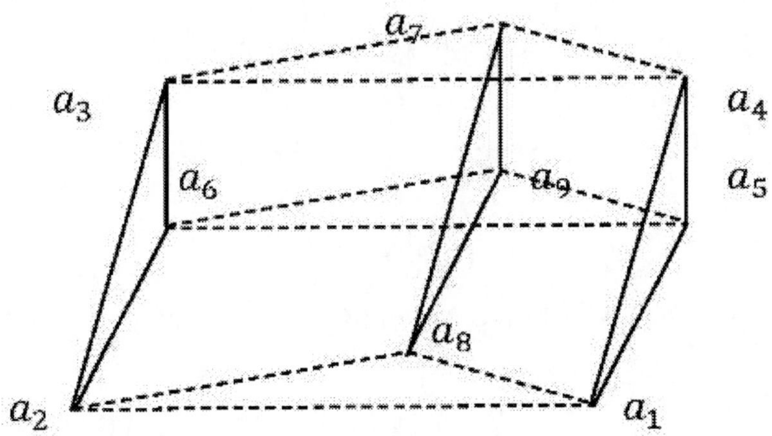

Figure 2.13. The of triangular prismahedron.

The components of the triangular prismahedron of the six triangular prisms are shown in Figure 2.14 (1-6).

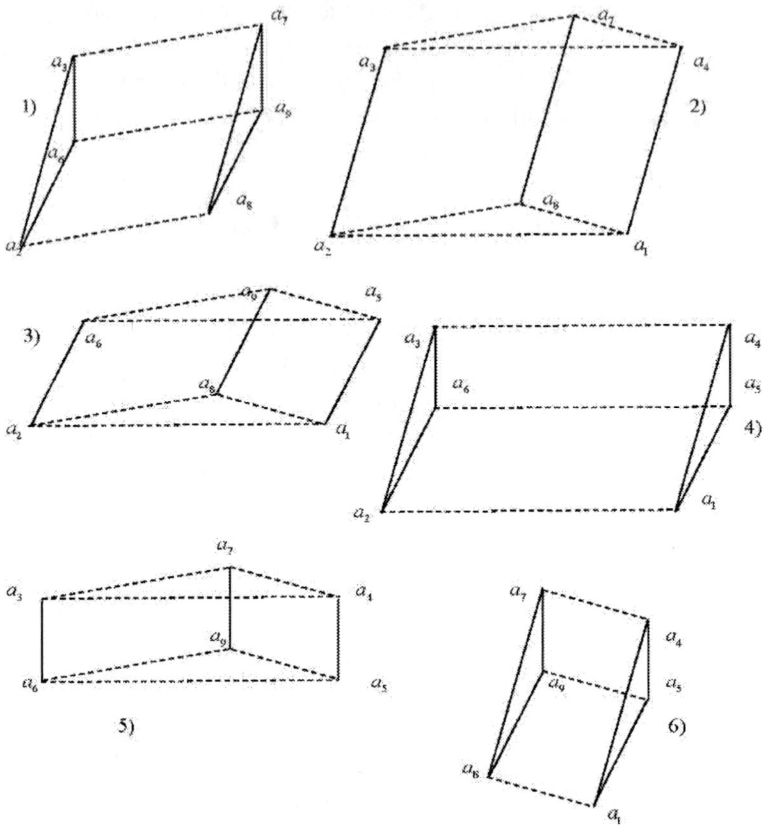

Figure 2.14. Facets of a triangular prismahedron.

1) $a_3a_2a_6a_7a_8a_9$, 2) $a_1a_2a_3a_4a_8a_8$, 3) $a_1a_2a_5a_6a_8a_9$, 4) $a_1a_2a_3a_4a_6a_5$
5) $a_3a_4a_6a_7a_5a_9$, 6) $a_1a_4a_5a_7a_8a_9$

Since the triangular prismahedron has dimension 4, its facets must adjoin each other along two -dimensional faces. Since a triangular prismahedron is not a regular figure, the common two-dimensional faces of triangular prismahedrons have two different types: triangles and quadrangles. It is possible to accurately enumerate the common two-dimensional faces of pairs of triangular prismahedrons:

1-2: $a_3a_2a_7a_8$, 1-4: $a_3a_2a_6$, 1-6: $a_9a_7a_8$, 1-5: $a_3a_6a_7a_9$ $a_6a_2a_9a_8$, 2-4: $a_1a_2a_3a_4$

2-3: $a_1a_2a_8$, 2-5: $a_3a_4a_7$, 2-6: $a_1a_4a_7a_8$, 3-4: $a_1a_2a_5a_6$, 3-5: $a_6a_9a_5$ 3-6: $a_1a_5a_9a_8$

4-5: $a_3a_4a_5a_6$, 4-6: $a_1a_4a_5$, 5-6: $a_4a_5a_7a_9$

To construct a polytope dual to a triangular prism, it is necessary to connect the center of each triangular prism by edges with the centers of other triangular prisms. These edges are images of triangles and quadrangles that are common in a pair of triangular prisms. Since the edges correspond to triangles and quadrangles, let us agree to denote the edges corresponding to triangles by a dotted line, and the edges corresponding to quadrilaterals by a solid line. Then the polytope dual to the triangular prismahedron has the form shown in Figure 2.15.

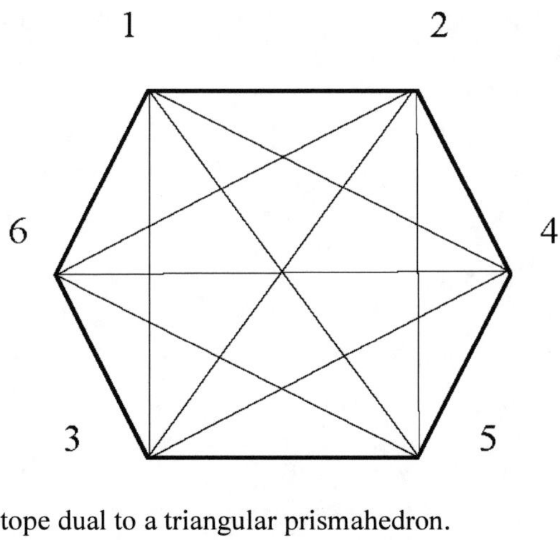

Figure 2.15. Polytope dual to a triangular prismahedron.

In the polytope in Figure 2.15, each vertex is connected by an edge to the rest of the vertices, i.e., is a simplex in accordance with the definition of a simplex (Zhizhin, 2019). Since the polytope in Figure 2.15 has 6 vertices, the dimension of the simplex is 5, in accordance with the definition of a simplex. The number of elements of various dimensions in a simplex is determined (Zhizhin, 2019) by the formula:

$$f_i(n) = C_{n+1}^{i+1}, \quad (2.7)$$

i-dimension of an element, n-dimension of the polytope.

By formula (2.7), we determine that the polytope in Figure 2.15 contains 15 edges and 15 three-dimensional polytopes:

$$f_1(5) = C_6^2 = 15, \quad f_3(5) = C_6^4 = 15$$

The edges: 61, 12, 24, 45, 35, 63, 13, 14, 34, 62, 25, 65, 15, 32, 64.

Since a simplex is a simplicial polytope, all its three-dimensional faces are tetrahedrons. Using Figure 2.15, we list all of its tetrahedrons:

$$T_1 = 1245, T_2 = 1365, T_3 = 3245, T_4 = 6231, T_5 = 1624, T_6 = 4365, T_7 = 1453, T_8 = 2635, T_9 = 6143,$$
$$T_{10} = 1256, T_{11} = 1243, T_{12} = 6245, T_{13} = 1234, T_{14} = 6243, T_{15} = 1645.$$

For what follows, it is useful to make sure that the polytope in Figure 2.15 satisfies the Euler -Poincaré equation and therefore exists. In addition to the number of vertices, edges and tetrahedron for substitution in the Euler-Poincaré equation, it is necessary to recalculate the number of two-dimensional faces and the number of four-dimensional faces according to Figure 2.15. From Figure 2.15 it follows that the polytope includes 20 triangles: 126, 124, 145, 135, 136, 134, 236, 235, 245, 265, 463, 435, 563, 564, 132, 162, 234, 462, 632, 165. This number coincides with the number determined by the formula (2.7) $f_2(5) = C_6^3 = 20$
follows that the polytope includes 6 polytopes 4-simplex: 61245, 12453, 24536, 45361, 53612, 36124. This number coincides with the number determined by the formula (2.7) $f_4(5) = C_6^5 = 6$

Substituting the obtained quantities $f_i(n), i = 0 \div 4,$
different dimensions into the Euler-Poincare equation (2.6), can obtain

$$6\text{-}15 + 20\text{-}15 + 4 = 2.$$

This means that equation (2.6) is satisfied for $n = 5$. Consequently, the polytope on Figure 2.15 has dimension 5.

Let's enumerate the edges and define their belonging to the listed the tetrahedrons.

$$61 \in T_2, T_4, T_5, T_9, T_{10}, T_{15}; 12 \in T_1, T_4, T_5, T_{10}, T_{11}, T_{13}; 24 \in T_1, T_5, T_3, T_{11}, T_{12}, T_{13}, T_{14}; 45 \in T_1, T_3, T_6, T_7, T_{12}, T_{15};$$
$$35 \in T_2, T_3, T_6, T_7, T_8; 63 \in T_6, T_4, T_2, T_8, T_9, T_{14}; 13 \in T_7, T_2, T_4, T_9, T_{11}, T_{13}; 14 \in T_1, T_5, T_7, T_9, T_{11}, T_{13}, T_{15};$$
$$34 \in T_3, T_6, T_7, T_9, T_{11}, T_{13}, T_{14}; 62 \in T_4, T_5, T_8, T_{10}, T_{12}, T_{14}; 25 \in T_1, T_3, T_8, T_{10}, T_{12}; 65 \in T_2, T_6, T_8, T_{10}, T_{12}, T_{15};$$
$$15 \in T_1, T_2, T_7, T_{10}, T_{15}; 32 \in T_3, T_4, T_8, T_{11}, T_{13}, T_{14}; 64 \in T_5, T_6, T_{12}, T_9, T_{14}, T_{15}.$$

Thus, there is a wonderful fact: the edges have different values of incidence to tetrahedrons. The edges 15, 35, 25 have value of incidence 5; the edges 61, 12, 13, 45, 35, 63, 62, 64, 32 have value of incidence 6; the edges 24, 14, 34 have value of incidence 7. It occurs when the triangular prismahedron has all edges with equal values of incidence. This is especially surprising, since in a simplex of dimension 4, all elements have correspondingly the same incidence coefficients. This implies a very important conclusion that an increase in the dimension of the polytope leads to a violation of the homogeneity of elements of different dimensions, taking into account their dimension. It should be noted that the fact that polytope 5 is a simplex in this case was obtained as a result of constructing a polytope dual with respect to a triangular prismahedron does not change the result, which has a general meaning. This once again indicates that an increase in the dimension of figures can lead to unforeseen consequences that are fundamentally different from figures in three-dimensional Euclidean space.

It should also be noted an interesting fact about the change in the dimension of the higher -dimensional polytope when constructing a polytope dual to it.

Conclusion

Geometric constructions of polytopes of higher dimensions of various types are considered: well-known polytopes of the n-cube, n-simplex, n-cross-polytopes type, as well as polytopes of a new type such as polytopic prisms, polytopic prismahedrons, polyincidental polytopes and polytopes dual to them. In this variety of different types of polytopes of the highest dimension, the question of whether geometric elements of a certain dimension belong to geometric elements of another dimension was investigated. It is proved that the coefficient incidence of an element of dimension i of the boundary complex of an n-cube and n-simplex is equal to the difference of the dimension of the cube or simplex n and the dimension of this element i. The construction of n-cross-polytopes of various dimensions, including large dimension, revealed that the incidence coefficient of elements of this polytope is substantially higher than the incidence coefficient of the elements of the n-cubes and n-simplexes for equal values of the dimension of polytopes. The incidence coefficients of all three types of polytopes increase linearly with increasing n, but the proportionality coefficient of this dependence in cross polytopes is twice as large as the proportionality coefficient of cubes and simplexes.

Significantly larger values of the incidence coefficients of the elements of the boundary complex of n-cross-polytopes as compared to the n-cubes and n-simplexes should be taken into account in the study of structures involving the n-cross-polytopes.

The mathematical representation of the polytope boundary complex of higher dimension corresponds to the physical representation of the surface (Adam, 1938) molecular formation (cluster). All atoms of molecular formation are in the boundary complex of polytope. Moreover, two types of the polytope higher dimension, corresponding to molecular formation, are possible. The first type contains within itself a region of internal points free of atoms. The second type does not have internal points; the whole body of the polytope is a boundary complex. The region of interior points can be used to place atoms in it from the external environment. In this case, the polytope changes the type from the first to the second. All chemical interactions can occur with atoms of the boundary complex, in principle, in any place of their location.

The values of the incidence coefficients of geometric elements in the products of polytopes (polytopic prisms and polytopic prismahedrons) often obey equation (2.2). The difference is observed if the composition of polytopes includes pyramids and cross-polytopes.

It has been established, that the polytopic prismahedrons have, in the general case, another surprising property: polytopes dual to the polytopic prismahedrons have faces (vertices, edges) with different incidence values with respect to higher dimensional faces in this polytope. Such polytopes form a new class: a class of poly-incident polytopes, which are obviously incorrect, but have certain symmetry elements. Analytical analysis of various types of polytopes, dual to the polytopic prismahedrons, is carried out.

For the first time, analytical laws for the dependence of the incidence coefficients of the elements of the n-cubes, the n-simplexes, and the n-cross-polytopes on the dimension of the polytope are established. According to these laws, the incidence coefficients increase linearly with increasing *n*, especially this increase occurs in the *n*-cross-polytopes.

Earlier, in the work of the author (Zhizhin, 2019), it was established that the homogeneity of higher-dimensional polytopes changes when constructing dual polytopes for them. This was shown by examples of polytopic prismahedrons. In this work, it is shown for the first time that an increase in the dimension of figures in itself leads to a violation of the uniformity of figures, i.e., to the appearance of geometric elements in a figure with different incidence values in relation to elements of a different dimension within this figure. This once again emphasizes the incorrectness of the distribution of the properties of geometric figures of a given dimension when the dimension of the figures is increased.

Chapter 3

Non-Euclidean Properties of the Geometry of Polytopes of Higher Dimension

Abstract

The inertial motions of a material point in a higher-dimensional space, which, in accordance with the geometric representations of Riemann, is considered closed and finite, are considered. The curvature of space is considered constant and equal to zero. It is shown that inertial motions in the local approximation in such *n*-dimensional geometry lead either to impulsive motions in a space of dimension *n*, or to sudden transitions of a material point from one space of dimension less than *n* to other spaces of dimension less than *n*, without breaking the continuity of the space of higher dimension *n*. It is shown that the stationary Schrödinger equation describing the distribution of electrons in the vicinity of the atomic nucleus has a solution, in principle, for any dimensionality of the space around the nucleus. As an example, a solution of the Schrödinger equation in a five-dimensional space is obtained. It is shown that the solution of the Schrödinger equation in *p*-dimensional space has *p* quantum numbers: the principal quantum number, the orbital quantum number and *p*-2 magnetic quantum numbers. Taking into account the spin quantum number, the total number of quantum numbers in *p*-dimensional space is $p + 1$. This leads to the possibility of increasing the number of quantum cells of orbitals and, consequently, to the possibility of increasing the valence of the elements.

Keywords: higher-dimensional space, quantum numbers, Euclidean geometry, non-Euclidean geometry, polytope, inertial motion, incidence

Introduction

After describing the features of the Euclidean and non-Euclidean geometries (Chapter 1) and the geometric properties of higher-dimensional polytopes (Chapter 2), it is necessary to compare the requirements of the *n*-dimensional geometry of Euclidean and the geometry of higher-dimensional polytopes.

The first part of the third chapter is devoted to this question. It formulates in detail the axioms of the connection, which are necessary in the n-dimensional geometry of Euclidean, and they are compared with the laws of incidence in the geometry of polytopes of the highest dimension. It is shown that the axioms of connection of the n-dimensional geometry of Euclidean do not agree with the laws of incidents in the geometry of polytopes of higher dimension. In addition, the axiom of parallelism of the geometry of Euclidean also cannot be applied in the geometry of polytopes of higher dimension, since the axiom of parallelism is based on the concept of infinity of straight lines, and in the geometry of polytopes of higher dimension, the concept of infinity is absent due to the initial concept of boundedness of polytopes of higher dimension. As you know, the geometry of Euclidean became the basis of classical mechanics. In this regard, in this chapter it is shown that the concepts of classical mechanics are also inapplicable for a polytope of higher dimension. The inertial motion of a material point in polytopes of the highest dimension is considered in detail. Since the faces of the polytope of dimension n are geometric elements with dimensions from 0 to n-1, so the inertial motion of a material point lead to a continuous transition of a material point from the space of one geometric element to the space of another geometric element, i.e., from one space to another space without breaking boundaries. Since it was previously proved that the molecules of chemical compounds, as a rule, have a higher dimension, this chapter discusses the question of the possible movement of electrons in the vicinity of the nucleus of an atom in a space of higher dimension. It is proved that the Schrödinger equation describing the motion of an electron in the vicinity of the nucleus of an atom has a solution in the case when the space in the vicinity of the nucleus has the highest dimension. This solution is given.

3.1. Axioms of Multidimensional Euclidean Geometry

Following Hilbert (Gilbert, 1930; Efimov, 1945), when describing the multidimensional Euclidean space, n systems of things are introduced. Objects of the first system are called points and are designated by the letters $A_1, A_2, A_3, \ldots$; objects of the second system are called straight lines and are denoted by the letters $a_1, a_2, a_3, \ldots$

dimensional planes and are designated $\alpha_2, \beta_2, \gamma_2, \ldots$

system are called three-dimensional planes and are denoted by $\alpha_3, \beta_3, \gamma_3, \ldots$ finally, the objects of the n-th system are called (n-1)-dimensional planes (hyperplanes) and are denoted by $\alpha_{n-1}, \beta_{n-1}, \gamma_{n-1}, \ldots$

There are some relationships between objects, which we express with the words: "lies on," "passes through," "between," etc. The entire collection of objects is called an n-dimensional Euclidean space if these relations satisfy the pipelines of five groups: Group 1 (connection axioms), Group 2 (order axioms), Group 3 (congruence axioms), Group 4 (continuity axioms), Group 5 (axioms of parallelism).

For what follows, in order to compare with the geometry of convex polytopes of higher dimension, the Group 1, which contains 8 axioms, is of paramount importance.

3.1.1. Connection Axioms

1. Whatever the two points A, B, there is a straight line a associated with each of these points.
2. Whatever the two different points A, B, there is at most one straight line associated with each of these points.
3. There are at least two points on every line. There are at least three points that are not collinear.
4. There is at least one point on each plane. Whatever the three points A_1, A_2, A_3, that do not lie on one straight line, there is a two-dimensional plane passing through the points A_1, A_2, A_3
   least four points that do not lie in the same two-dimensional plane.
   Whatever the four points A_1, A_2, A_3, A_4
   two-dimensional plane, there is a three-dimensional plane passing through the points A_1, A_2, A_3, A_4
   not lie in the same three-dimensional plane.
   Whatever the n points $A_1, A_2, \ldots, A_n$ n-2)-dimensional plane, there is an (n-1)-dimensional plane (hyperplane) passing through the points $A_1, A_2, \ldots, A_n$
5. Whatever the three points $A_1, A_2, A_3,$
   line, there is no more than one two-dimensional plane passing through the points $A_1, A_2, A_3,$ k + 1 points

$A_1, A_2, ..., A_{k+1}$ that do not lie in one (k-1)-dimensional plane (k =3, 4, …, n – 1), there is at most one k-dimensional plane passing through the points $A_1, A_2, ..., A_{k+1}$

6. If k + 1 points of the k-dimensional plane α_k, that do not lie on one (k-1)-dimensional plane, lie on the l-dimensional plane α_l,, then each point of the plane α_k, lies on α_l

   α_k lies in the plane α_l
7. If the k-dimensional plane α_k l-dimensional plane $\alpha_l, k + l - n = m > 0$, have a common point A, then they also have m common points $A_1, A_2, ..., A_m$

   $A, A_1, A_2, ..., A_m$ does not belong to one (m-1)-dimensional plane.
8. There are n + 1 points that do not lie in the same hyperplane.

Further, it is necessary to consistently formulate for n-dimensional Euclidean space the order axiom, the congruence axiom, the continuity axiom, and the parallelism axiom. However, these axioms can be taken without change from elementary geometry for three-dimensional space (Gilbert, 1930; Efimov, 1945). In particular, the axiom of parallelism has the same form: Let a be an arbitrary line and A be a point lying outside the line a; then in the plane defined by point A and straight-line a, you can draw at most one straight line passing through A and disjoint a. It is only necessary, wherever in elementary geometry we are talking about a plane, to mean a two -dimensional plane.

With the help of these axioms, it turns out to be possible to prove for objects of any three -dimensional plane all theorems of elementary geometry, and for objects of planes of higher dimension and the entire space, their corresponding generalizations.

3.2. About the Impossibility of the Axiom Systems of the N-Dimensional Geometry of Euclides for Higher Dimensional Popytopes

If we take, following Hilbert, elements of different dimensions of a polytope of higher dimension as a system of things of different dimensions, then we can see that the axioms of Euclid's n -dimensional geometry leads to a contradiction with the geometry of polytopes of higher dimension. For example, in link axiom 4, the condition is set: "Whatever the three points A_1, A_2, A_3, that do not lie on one straight line, there is a two-dimensional plane passing through the points A_1, A_2, A_3 "

For a polytope structure of higher dimension, which by definition has limited dimensions in all coordinates, this means that no matter what three vertices of the polytope A_1, A_2, A_3

on one edge (one-dimensional element), there is a two -dimensional face incident to three vertices A_1, A_2, A_3

dimensional polytope, considered in the previous section, this does not hold. Three vertices of a polytope can only in special cases be incident to one two-dimensional face. In general, this is not the case. In general, the laws of incidence in the polytope of higher dimension have no analogues in the proposed system of Hilbert's axioms for the n-dimensional geometry of Euclid. For example, further in the axiom 4 it is stated: "Whatever the four points A_1, A_2, A_3, A_4

is a three-dimensional plane passing through the points A_1, A_2, A_3, A_4 " For the geometry of a polytope of the highest dimension, this would mean that for any four vertices A_1, A_2, A_3, A_4

face, there is a three-dimensional face incident to these vertices. This is also a completely wrong statement for the geometry of a higher-dimensional polytope. Similar contradictions are observed for all eight connection axioms of the n-dimensional geometry of Euclid.

Consequently, the system of constraint axioms necessary for the n-dimensional geometry of Euclid is inapplicable when considering systems of elements of different dimensions in polytopes of higher dimensions.

If we consider the points of the system to be the possible points of the polytope of the highest dimension (and not just the vertices), consider all the possible segments connecting the selected points as one-dimensional elements, then here we again come to contradictions with the system of axioms of the n-dimensional geometry of Euclid. The fact is that the space of a polytope of higher dimension n, as follows from the previous section, consists of faces of different dimensions from 0 to $n - 1$, and a set of interior points for the polytope, which has dimension n. Moreover, the set of internal points, as it was proved in the previous section, does not always exist. The line segments connecting points inside a certain face are located naturally inside the face. Moreover, the segments are not straight since their sizes are limited. The sections of the faces are also limited, so they cannot be considered as infinite planes of a certain dimension. In this regard, the axiom of parallelism of the system of axioms of the n-dimensional geometry of Euclid becomes inapplicable since it is based on the definition of parallel lines of infinite length.

Thus, the geometry of polytopes of the highest dimension should be considered non-Euclidean. This non-Euclidean geometry is not a consequence of the curvature of space, as in the geometries of Lobachevsky and Riemann.

Space in mathematics is a logically conceivable form (or structure) that serves as a medium in which other forms or constructions exist (Encyclopedia of Mathematics, 1977). In this definition, the essential meaning is that there is a logically conceivable form. The visual image of any object on the retina is two-dimensional. Therefore, any object is initially perceived as two-dimensional. Likewise, we also perceive only the surface of the object, i.e., again the tangible image of the object is two-dimensional. The idea of the three-dimensionality of an object is formed only as a result of comparing mismatched images in the right and left eyes. Comparison is the result of thinking, as a result of which judgment appears. Therefore, the idea of even the three-dimensionality of objects is an abstract idea and, because of this, it is difficult for many people. Let abstraction leads us to conceptual space (the space studied by geometry-geometric space (Poincaré, 1895). But this conceptual space arose on the basis of the perception of space, i.e., based on perceptual space (representation space). And the perceptual space itself is a reflection of the real material space. Perceptual space is an image of that real space that exists regardless of whether we perceive it or not, and from what kind of perception we receive from it. However, given that perceptual space is a consequence of mental activity, the image of real space is ambiguous. It can take into account various ideas of a person about the environment, as well as certain properties of the surrounding world, which seem to be the most important to a person in given specific circumstances. This is also the reason for the existence of different geometries. It should be borne in mind here that the geometric axioms "are neither synthetic a priori judgments, nor experimental facts. They are conditional provisions (agreements): when choosing between all possible agreements, we are guided by experimental facts, but the choice remains free and is limited only by the need to avoid any contradiction." (Poincaré, 1895). Historically, two main ideas about space have developed, within the framework of which modifications of these representations and the corresponding geometries have arisen. According to the first of these concepts, associated with the names of Aristotle and Leibniz, real space is a property of the position of material objects, i.e., it is inextricably linked with matter. The development of these ideas led to the well-known position of philosophy that there is no space without matter, just as there is no matter without space, and that space is a form of existence of matter. According to the second of these concepts, associated with the names of Democritus and Newton, space is the receptacle of all material objects that have no effect on space (Einstein, 1930). It is this concept of space that was the philosophical basis of Euclid's geometry. According to this concept, geometric space is continuous, infinite, three-dimensional, homogeneous, isotropic. Lobachevsky of consistent non-Euclidean geometry marked the beginning of the rapid development of geometry (Lobachevsky, 1835). Work on non-Euclidean geometry was continued by Riemann. He introduced the concept of n-extended manifolds (Riemann, 1854). Riemann showed that the manifold of $(n + 1)$ dimensions is the manifold of the 1st dimension, consisting of elements (points) and dimensions. Riemann introduced a measure definition, according to which the distance from any point to the nearest point is approximately expressed by the same expression as in Euclidean geometry. The difference between Riemann's measure definition is that this expression is approximate. This implies the concept of the curvature of space. The curvature of space, which Riemann considered, is positive zero. Later this space was called the Riemannian space (Klein, 1935; Aleksandrov, 1956; Cartan, 2010; Gilbert Conn -Vossen, 1932). While Riemann's work appeared in science there was a great interest in the theory of relativity. Riemann's geometry has been applied in the theory of relativity. However, in Riemann's work, in addition to the concept of space with positive curvature, there are fundamental points not related to the theory of relativity that require analysis at the present time. First, Riemann notes that examples of multidimensional manifolds can be manifolds formed by the contours of geometric shapes. Secondly, when defining an n-extended manifold, Riemann did not use the concept of infinity of space. Moreover, Riemann admits that space is finite. An example of such spaces can be spaces of any small positive curvature. But there can be, as follows from Riemann's work, and other examples, in particular manifolds formed by the contours of geometric figures.

At the time when Riemann wrote his work, the concept of multidimensional geometric figures was not developed. It is significant that, as it was established later (Grünbaum, 1967), geometric figures of the highest dimension have a complex structure, consisting of a boundary complex, which includes geometric elements of various dimensions. In addition, for the existence of a geometric figure of higher dimension (polytope), it is necessary to fulfill a certain relation between the numbers of elements of different dimensions – the Euler-Poincaré equation (Poincaré, 1895), obtained by Poincaré as an extension of the Euler equation for polytopes of higher dimension. Many mathematicians did not take this relation into account and obtained incorrect results when trying to construct polytopes of higher dimension (Minkovskii, 1911; Delone, 1929, 1937, 1961; Venkov, 1959).

Taking into account that, according to Riemann's ideas, the space of higher dimension is closed and cannot be infinite, we come to the need to study polytopes of higher dimension as closed spaces of higher dimension. The monographs (Zhizhin, 2018; 2019 a, b; 2021 a, b) convincingly show that almost all chemical and biological compounds are objects of the highest dimension. Therefore, the question of analyzing the features of at least the simplest inertial motion in these objects is ripe. This work is devoted to this.

3.3. Inappropriatness of the Basic Principles of Classical Mechanics for the Analysis of Motion in a Space of Higher Dimensions

The fundamental difference between the space of higher dimension and Euclidean space, which is the basis of classical mechanics, is the limited space of higher dimension. This space has a boundary complex of geometric elements of various dimensions. Therefore, the space of the highest dimension can be neither infinite, nor homogeneous, nor isotropic. Isotropy and homogeneity are violated by the peculiarities of the shape of polytopes of higher dimensions, which in the general case have different types of vertices and their neighborhoods. Due to the limited space of the highest dimension, the law of inertia of classical mechanics cannot be applied to it: a body cannot move uniformly and rectilinearly in the absence of forces acting on it to infinity, it will certainly reach a boundary complex. The law of inertia in the space of the highest dimension can be accepted only in a local approximation far from the boundary complex.

Behavior of a material point after reaching the boundary complex of the space of the highest dimension of a hypothetical analysis of possible variants of this behavior.

3.3.1. Movement of a Material Point in Four – Dimensional Spaces

3.3.1.1. Movement of a Material Point in a 4-Cube

Rust there is a four-dimensional cube (4-cube), Figure 3.1.

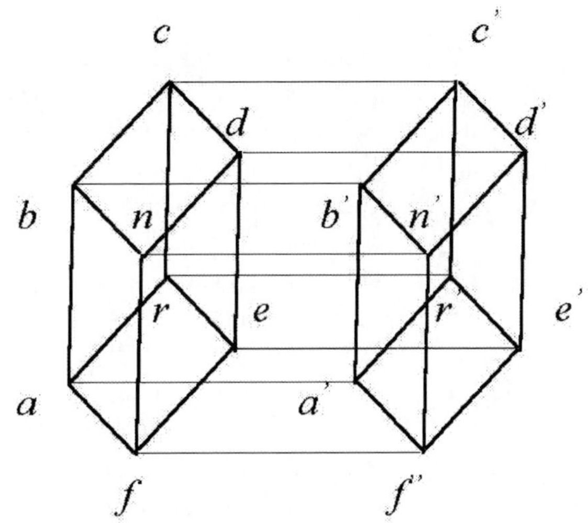

Figure 3.1. The four-dimensional cube.

If we place the origin of coordinates $(x=0, y=0, z=0, t=0)$ center of the cube, then the set of internal points of the 4-cube is determined by a set of inequalities

$$-\frac{1}{2}<x<\frac{1}{2}, -\frac{1}{2}<y<\frac{1}{2}, -\frac{1}{2}<z<\frac{1}{2}, -\frac{1}{2}<t<\frac{1}{2}.$$

Boundary complex 4 – cube (n = 4) includes $2^4 = 16$ vertices, $2^3 \cdot 4 = 32$ edges, $2^2 \cdot 6 = 24$ flat faces, $2 \cdot 4 = 8$

$$abcdrenf, a'b'c'd'r'e'n'f', bcdnb'c'd'n', arefa'r'e'f',$$
$$cdrec'd'r'e', bnafb'n'a'f', abcra'b'c'r', fndef'n'd'e'.$$

Incidence coefficients of elements of different dimensions are significant (Zhizhin, 2019a)

$$k_{b(i=0)}(4) = 4, k_{b(i=1)}(4) = 3, k_{b(i=2)}(4) = 2, k_{b(i=3)}(4) = 1. \quad (3.1)$$

Suppose that some material point A belongs to the set of interior points of the 4-cube. If an external force does not act on it, then according to the local law of inertia, it can be at rest or in rectilinear uniform motion in one of the directions in the 4-cube. It will be in this motion until it approaches one of the flat faces of one of the eight three-dimensional cubes included in the boundary complex.

Here it can be reflected from this face and, according to the law of conservation of momentum (the angle of incidence is equal to the angle of reflection), move to one of the flat faces of another three-dimensional cube of the boundary complex. This movement of a point can continue, successively reflecting off flat faces, remaining in a four-dimensional set of interior points. In addition, a material point A with sufficient energy of motion can cross the oncoming flat face of the nearest three-dimensional cube, overcoming the possible surface energy of the face, since material bodies are located at the vertices of the face. Here it should be taken into account that, in accordance with the incidence coefficients (3.1) in a four-dimensional cube, each two-dimensional face is common to two three-dimensional faces. In the future, the material point A will move from one three-dimensional space (three-dimensional cube) to another three-dimensional space, sequentially crossing the flat faces of these cubes. This process can continue indefinitely. If point A reaches a flat face, which is the outer boundary of a four-dimensional cube, then point A will reflect from this boundary and continue to move through space in three-dimensional cubes, since a four-dimensional cube is a closed limited space. From the point of view of an external observer, point A suddenly disappears from one three-dimensional space and appears in another three-dimensional space without breaking the border between these spaces.

3.3.1.2. Movement of a Material Point in a 4-Simplex

Let there be a four-dimensional space in the form of a 4 – simplex (Figure 3.2).

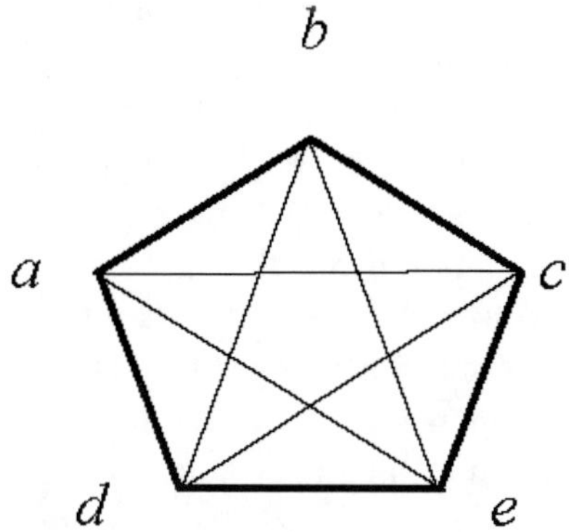

Figure 3.2. The 4 – simplex.

The number of elements of different dimensions in the n-simplex is determined by the equation (Zhizhin, 2013, 2014a)

$$f_i(n) = C_{n+1}^{i+1}, i = 0,1,2,...,n. \tag{3.2}$$

According to equation (3.2) for the n – simplex for dimension 4, we have $f_0(4) = 5, f_1(4) = 10, f_2(4) = 10, f_3(4) = 5.$ – simplex has 5 vertices, 10 edges, 10 flat faces (triangles), 5 tetrahedrons

abde, acde, abcd, abce, ebcd.

It follows from Figure 3.2 that the incidence coefficients of the elements of the boundary complex the 4 – simplex have the values

$$k_{b(i=0)}(4) = 4, k_{b(i=1)}(4) = 3, k_{b(i=2)}(4) = 2, k_{b(i=3)}(4) = 1.$$

Points that belong to the 4-simplex, but are not included in any of the tetrahedron, are internal points. Let one of the points be point A and external forces do not act on it. Then it can move rectilinearly and uniformly along one of the directions in the 4-simplex.

Let's say point A is in motion. Then sooner or later it will reach the boundary complex, i.e., one of the three-dimensional facets. After meeting with the boundary complex, point A will either be reflected from one of the flat faces of the facet, preserving the law of momentum (the angle of incidence is equal to the angle of reflection), and further will continue to move sequentially reflecting from the flat faces of the facets. Or point A, with sufficient energy of motion, overcoming the surface energy of the flat face, will enter the three-dimensional space of this facet. Therefore, point A will move from the four-dimensional space to the three-dimensional space of this facet. Further movement of the material point A will be accompanied by transitions from one three-dimensional facet to other three-dimensional facets, depending on the initial direction of movement. For an external observer, this movement will be presented as a sudden disappearance of point A from one of the three-dimensional spaces and the successive appearance of this point in other three-dimensional spaces without violating the integrity of the boundary complex.

3.3.1.3. Movement of a Material Point in a 4-Cross-Polytope

Let there be a four-dimensional space in the form of a 4 – cross-polytope (Figure 3.3). The structure of the boundary complex of the n – cross-polytope differs significantly from the structure of the boundary complexes of the n-cube and n-simplex (Zhizhin, 2019a). The number of elements of different dimensions in the n – cross-polytope is determined by the equation (Zhizhin, 2013, 2014a)

$$f_i(n) = 2^{1+i} C_n^{n-1-i}. \tag{3.3}$$

According to equation (3.3) for the n – cross-polytope of the smallest possible dimension 4, we have $f_0(4) = 8, f_1(4) = 24, f_2(4) = 32, f_3(4) = 16.$ Thus, the 4 – cross – polytope has 8 vertices, 24 edges, 32 flat faces (triangles), 16 tetrahedrons

abcd, abch, bcde, cdef, gdef, efhg, afhg, abhg, adfg, behg, cfag, dgab, ebhc, afcd, bgde, hcef.

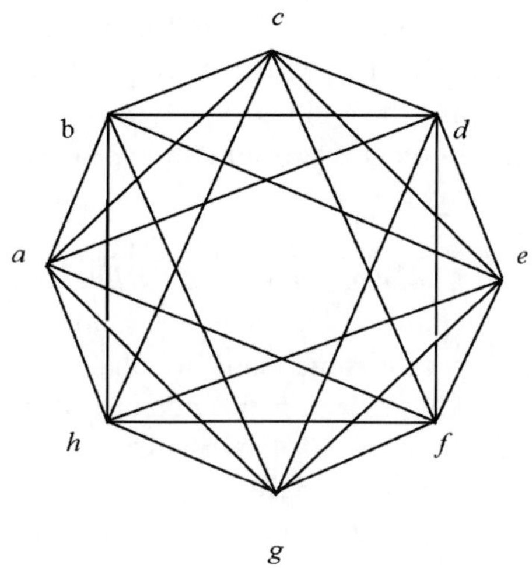

Figure 3.3. The 4 – cross-polytope.

It is essential that the vertices *n*-cross-polytope located opposite to each other relative to the center do not have an edge connection. It follows from Figure 3.3 that the incidence coefficients of the elements of the boundary complex the 4 – cross-polytope have the values

$$k_{b(i=0)}(4)=6, k_{b(i=1)}(4)=4, k_{b(i=2)}(4)=2, k_{b(i=3)}(4)=1.$$

Points that belong to the 4 – cross-polytope, but are not included in any of the tetrahedra, are internal points. Let one of the points be point A and external forces do not act on it. Then it can move rectilinearly and uniformly along one of the directions in the 4 – cross-polytope.

Let's say point A is in motion. Then sooner or later it will reach the boundary complex, i.e., one of the three-dimensional facets. After meeting with the boundary complex, point A will either be reflected from one of the flat faces of the facet, preserving the law of momentum (the angle of incidence is equal to the angle of reflection), and further will continue to move sequentially reflecting from the flat faces of the facets. Or point A, with sufficient energy of motion, overcoming the surface energy of the flat face, will enter the three-dimensional space of this facet. Therefore, point A will move from the four-dimensional space to the three-dimensional space of this facet. Further movement of the material point A will be accompanied by transitions from one three-dimensional facet to other three-dimensional facets, depending on the initial direction of movement. For an external observer, this movement will be presented as a sudden disappearance of point A from one of the three-dimensional spaces and the successive appearance of this point in other three-dimensional spaces without violating the integrity of the boundary complex.

3.3.1.4. Movement of a Material Point in n – Dimensional Spaces (n > 4)

If the dimension of the space is more than four, then in accordance with the laws of incidence (Zhizhin, 2019 a), the number of spaces of a certain dimension i incident to the space of dimension i-1 increases. In particular, the number of three-dimensional spaces incident to one two-dimensional face increases in proportion to the dimension of the general space n. In this connection, the nature of the inertial motion of a material point in this space is significantly complicated due to the increase in the options for the transition of a material point from some space to others, although the general nature of these movements remains: reflection from the outer boundary of space and the intersection of the interfaces between the spaces.

3.4. On the Possibile Electronic Structure of Atoms in a Space of Higher Dimension

The studies carried out in the monograph of G. Zhizhin "Chemical Compound Structures and Higher Dimension of Molecules: Emerging Research and Opportunities" (Zhizhin, 2018) convincingly showed that many elements of the periodic system of Mendeleev elements form molecules whose spatial structure can be geometrically represented as a convex polytope. Moreover, the dimension of this polytope, determined by the Euler-Poincaré relation between the geometric elements forming the polytope (Poincare, 1895), is usually higher than three. In addition, in this monograph it is noted that most elements of the periodic system exhibit valence in compounds higher than the valence determined by the location of the element in the periodic table. It is known that the calculation of the valence of a chemical element based on its location in the periodic table is based on the electronic structure of atoms (s -, p -, d -, f – orbitals), determined by solving the Schrodinger equation (Schrodinger, 1927) under the assumption of the three-dimensionality of the space surrounding the nucleus of the atom. To explain the observed differences in the valence of atoms from valence, determined from the location of the element in the periodic table, researchers resorted to the construction of various models of the interaction of electrons in the vicinity of atomic nuclei. These models are quite diverse and still remain at the level of hypotheses. In this connection, this chapter analyzes the possibility of obtaining a solution of the Schrödinger equation in a space of higher dimension. Although before it was believed that the solutions of the Schrödinger equation in a space with a dimension greater than three do not exist (Büchel, 1963; Freeman, 1969; Gurevich & Mostepanenko, 1971).

3.4.1. The Stationary Schrödinger Equation in a *P*– Dimensional Metric Space

If a solution of the Schrödinger equation in the case of four-dimensional space was obtained in the monograph of Zhizhin (Zhizhin, 2018), in this book the solution of the Schrodinger equation for any finite dimensionality of space is obtained in principle.

It is shown that the solution of the Schrödinger equation in a space of higher dimension leads to the existence of a large number of quantum numbers and, consequently, to an increase in the number of quantum cells of atoms and an increase in their valence.

The stationary Schrödinger equation in a p-dimensional metric space has view

$$-\frac{\hbar}{2m}\nabla_p^2\Psi + U\Psi = E\Psi. \tag{3.4}$$

Where Ψ-the wave`s function, U-the potential energy of an electron (the function of radius r), E – the kinetic energy of electron, ∇_p^2- Laplacian in p-dimensional metric space, m – the mass of electron, $\hbar$ –Planck constant.

The potential energy $U(r)$ – this potential energy of the Coulomb attraction of an electron to the nucleus and can be written in explicit form

$$U(r) = -\frac{Ze^2}{r}. \tag{3.5}$$

Where eZ -ion`s charge, e – electron's charge.

It is essential that the form of the potential energy (3.5) does not depend on the dimensionality of space, and it is defined as a function inversely proportional to the radius in the first degree. Since in a space of any dimension, the potential energy is a function of the distance between the charges (of the electron and the nucleus). The incorrectness in previous studies of the solution of the Schrödinger equation in spaces of higher dimension is that the authors of these works (Büchel, 1963; Freeman, 1969; Gurevich & Mostepanenko, 1971) recorded potential energy without a physical basis as a function of the radius to the power of p-2. Such an entry goes to formula (3.5) only in three-dimensional space, but in a space of the high dimensional this record is not true and leads to an absurd conclusion about the impossibility of the existence of a discrete atomic structure in a space of higher dimension.

Equation (3.4), having transferred to the atomic system of units, taking into account (3.5), in the Cartesian coordinate system has the form

$$\frac{\partial^2\Psi}{\partial x_1^2}+\frac{\partial^2\Psi}{\partial x_2^2}+\ldots+\frac{\partial^2\Psi}{\partial x_p^2}+\left(\frac{c}{r}+2E\right)\Psi=0. \tag{3.6}$$

Where c – constant, $r=\sqrt{x_1^2+\ldots+x_p^2}$.

For further analysis, it is convenient to go over into a spherical coordinate system, where the radius of one of the p variables. The remaining p-1 variables are the angles between the vector of length r to the point of location of the particle and the unit vectors of the Cartesian system of coordinates.

It can be noted that Gurevich & Mostepanenko (1971) were confused with coordinate systems. For example, the velocity vector in the n-dimensional space has n + 1 components in their work. Babenko (2015), considering n-dimensional space, describes the one-dimensional Schrödinger equation, forgetting the remaining variables.

In an n-dimensional space, the analytic connection between the coordinates is expressed by equalities (Hobson, 1931).

$$\begin{aligned}
x_1 &= r\sin\alpha_1 \sin\alpha_2 ... \sin\alpha_{p-1},\\
x_2 &= r\cos\alpha_1 \sin\alpha_2 ... \sin\alpha_{p-1},\\
x_3 &= r\cos\alpha_2 \sin\alpha_3 ... \sin\alpha_{p-1},\\
&.\\
x_{p-1} &= r\cos\alpha_{p-2} \sin\alpha_{p-2},\\
x_p &= r\cos\alpha_{p-1}.
\end{aligned} \tag{3.7}$$

The Laplacian in the Schrodinger equation in a spherical coordinate system in p-dimensional space can be written in the form

$$\nabla_p^2 = \frac{1}{L_1 L_2 L_3 .. L_p}\left[\frac{\partial}{\partial r}\left(\frac{L_2 L_3 .. L_p}{L_1}\frac{\partial\Psi}{\partial r}\right) + \frac{\partial}{\partial\alpha_1}\left(\frac{L_1 L_3 .. L_p}{L_2}\frac{\partial\Psi}{\partial\alpha_1}\right) + \frac{\partial}{\partial\alpha_2}\left(\frac{L_1 L_2 L_4 ... L_p}{L_3}\frac{\partial\Psi}{\partial\alpha_2}\right) + \right.$$
$$\left. + \frac{\partial}{\partial\alpha_3}\left(\frac{L_1 L_2 L_3 L_5 ... L_p}{L_4}\frac{\partial\Psi}{\partial\alpha_3}\right) + ... + \frac{\partial}{\partial\alpha_{p-1}}\left(\frac{L_1 L_2 ... L_{p-1}}{L_p}\frac{\partial\Psi}{\partial\alpha_{p-1}}\right)\right]. \tag{3.8}$$

Where L_i $\qquad i = 1, 2, \ldots, p,$

$$\begin{aligned}
L_1 &= \sqrt{\left(\frac{\partial x_1}{\partial r}\right)^2 + \left(\frac{\partial x_2}{\partial r}\right)^2 + ... + \left(\frac{x_p}{\partial r}\right)^2},\\
L_2 &= \sqrt{\left(\frac{\partial x_1}{\partial\alpha_1}\right)^2 + \left(\frac{\partial x_2}{\partial\alpha_1}\right)^2 + ... + \left(\frac{x_p}{\partial\alpha_1}\right)^2},\\
&.\\
&.\\
L_p &= \sqrt{\left(\frac{\partial x_1}{\partial\alpha_{p-1}}\right)^2 + \left(\frac{\partial x_2}{\partial\alpha_{p-1}}\right)^2 + ... + \left(\frac{x_p}{\partial\alpha_{p-1}}\right)^2}.
\end{aligned} \tag{3.9}$$

Differentiating the equality (3.7) with respect to the independent variables, substituting the derivatives in the Lame coefficients (3.9), can obtain after the transformations

$$L_1 = 1, L_2 = r\sin\alpha_2 ... \sin\alpha_{p-1}, L_3 = r\sin\alpha_3 ... \sin\alpha_{p-1},$$
$$L_4 = r\sin\alpha_4 ... \sin\alpha_{p-1}, ..., L_{p-1} = r\sin\alpha_{p-1}, L_p = r. \tag{3.10}$$

The Schrödinger equation in a spherical coordinate system in *p*-dimensional space has view

$$\nabla_p^2\Psi + \left(\frac{2}{r} + 2E\right)\Psi = 0. \tag{3.11}$$

Where $\nabla_p^2\Psi$ – (3.10).

Then Schrödinger equation (3.11) has view

$$\frac{1}{r^{p-1}}\frac{\partial}{\partial r}\left(r^{p-1}\frac{\partial\Psi}{\partial r}\right) + \frac{1}{r^2\sin\alpha_2{}^2 ... \sin\alpha_{p-1}{}^2}\frac{\partial^2\Psi}{\partial\alpha_1{}^2} + \frac{1}{r^2\sin\alpha_2\sin\alpha_3{}^2 ... \sin^2\alpha_{p-1}}\left(\sin\alpha_2\frac{\partial\Psi}{\partial\alpha_2}\right) +$$
$$\frac{1}{r^2\sin\alpha_3{}^2\sin^2\alpha_4 ... \sin^2\alpha_{p-1}}\frac{\partial}{\partial\alpha_3}\left(\sin\alpha_3{}^2\frac{\partial\Psi}{\partial\alpha_3}\right) + ... + \frac{1}{r^2\sin^2\alpha_{p-1}}\frac{\partial}{\partial\alpha_{p-1}}\left(\sin^2\alpha_{p-1}\frac{\partial\Psi}{\partial\alpha_{p-1}}\right) +$$
$$+2\frac{\Psi}{r} + 2E\Psi = 0. \tag{3.12}$$

3.4.2. The Decision of the Stationary Schrödinger Equation in a P – Dimension Metric Space

A solution of equation (3.12) can be obtained by separating the variables assuming that

$$\Psi(r, \alpha_1, ..., \alpha_{p-1}) = R(r)\Phi_1(\alpha_1)\Phi_2(\alpha_2)...\Phi_{p-1}(\alpha_{p-1}). \tag{3.13}$$

We substitute (3.13) into (3.12)

$$\frac{1}{r^{p-3}R(r)}\frac{d}{dr}\left(r^{p-1}\frac{dR(r)}{dr}\right)+2r^{2}\left(\frac{1}{r}+E\right)=-\frac{1}{\Phi_1(\alpha_1)\sin\alpha_2^{\,2}...\sin\alpha_{p-1}^{\,2}}\frac{d^2\Phi_1(\alpha_1)}{d\alpha_1^{\,2}}-$$
$$-\frac{1}{\Phi_2(\alpha_2)\sin\alpha_2\sin\alpha_3^{\,2}...\sin^2\alpha_{p-1}}\frac{d}{d\alpha_2}\left(\sin\alpha_2\frac{d\Phi_2(\alpha_2)}{d\alpha_2}\right)-$$
$$-\frac{1}{\Phi_3(\alpha_3)\sin\alpha_3^{\,2}\sin^2\alpha_4...\sin^2\alpha_{p-1}}\frac{d}{d\alpha_3}\left(\sin\alpha_3^{\,2}\frac{d\Phi_3(\alpha_3)}{d\alpha_3}\right)-$$
$$-...-\frac{1}{\Phi_{p-1}\sin^2\alpha_{p-1}}\frac{d}{d\alpha_{p-1}}\left(\sin^2\alpha_{p-1}\frac{d\Phi_{p-1}}{d\alpha_{p-1}}\right).$$
(3.14)

The left part of (3.14) is depended at radius r, and rite part is dependent only at angels.

Therefore, both parts equality (3.14) equal to same constant C_1. From left part of (3.14) there is

$$\frac{1}{r^{p-3}R(r)}\frac{d}{dr}\left(r^{p-1}\frac{dR(r)}{dr}\right)+2r^{2}\left(\frac{1}{r}+E\right)=C_1.$$
(3.15)

From right part of (3.14) there is

$$C_1=-\frac{1}{\Phi_1(\alpha_1)\sin\alpha_2^{\,2}...\sin\alpha_{p-1}^{\,2}}\frac{d^2\Phi_1(\alpha_1)}{d\alpha_1^{\,2}}-$$
$$-\frac{1}{\Phi_2(\alpha_2)\sin\alpha_2\sin\alpha_3^{\,2}...\sin^2\alpha_{p-1}}\frac{d}{d\alpha_2}\left(\sin\alpha_2\frac{d\Phi_2(\alpha_2)}{d\alpha_2}\right)-$$
$$-\frac{1}{\Phi_3(\alpha_3)\sin\alpha_3^{\,2}...\sin^2{}_{p-1}}\frac{d}{d\alpha_3}\left(\sin\alpha_3^{\,2}\frac{d\Phi_3(\alpha_3)}{d\alpha_3}\right)-...-\frac{1}{\Phi_{p-1}\sin^2\alpha_{p-1}}\frac{d}{d\alpha_{p-1}}\left(\sin^2\alpha_{p-1}\frac{d\Phi_{p-1}}{d\alpha_{p-1}}\right).$$
(3.16)

The equation (3.14) can rewrite as follows

$$C_1\sin\alpha_{p-1}^{\,2}+\frac{1}{\Phi_{p-1}(\alpha_{p-1})}\frac{d}{d\alpha_{p-1}}\left(\sin\alpha_{p-1}^{\,2}\frac{d\Phi_{p-1}(\alpha_{p-1})}{d\alpha_{p-1}}\right)=-\frac{1}{\Phi_1(\alpha_1)\sin\alpha_2^{\,2}...\sin^2{}_{p-2}}\frac{d^2\Phi_1(\alpha_1)}{d\alpha_1^{\,2}}-$$
$$-\frac{1}{\Phi_2(\alpha_2)\sin\alpha_2\sin^2\alpha_3...\sin^2\alpha_{p-2}}\frac{d}{d\alpha_2}\left(\sin\alpha_2\frac{d\Phi_2(\alpha_2)}{d\alpha_2}\right)-$$
$$-\frac{1}{\Phi_3(\alpha_3)\sin^2\alpha_3...\sin^2\alpha_{p-2}}\frac{d}{d\alpha_3}\left(\sin^2\alpha_3\frac{d\Phi_3(\alpha_3)}{d\alpha_3}\right)-...$$
$$-\frac{1}{\Phi_{p-2}(\alpha_{p-2})\sin^2\alpha_{p-2}}\frac{d}{d\alpha_{p-2}}\left(\sin^2\alpha_{p-2}\frac{d\Phi_{p-2}(\alpha_{p-2})}{d\alpha_{p-2}}\right).$$
(3.17)

The left part of (3.17) is depended at angle α_{p-1} only, and the right part (3.17) depended at angles $\alpha_2,...,\alpha_{p-2}$

to same constant C_2

$$C_1 \sin\alpha_{p-1}{}^2 + \frac{1}{\Phi_{p-1}(\alpha_{p-1})}\frac{d}{d\alpha_{p-1}}\left(\sin\alpha_{p-1}{}^2 \frac{d\Phi_{p-1}(\alpha_{p-1})}{d\alpha_{p-1}}\right) = C_2. \tag{3.18}$$

From the right part (3.17) there is

$$-\frac{1}{\Phi_1(\alpha_1)}\frac{d^2\Phi_1(\alpha_1)}{d\alpha_1^2} = C_2\sin^2\alpha_2...\sin^2\alpha_{p-2} + \frac{\sin\alpha_2}{\Phi_2(\alpha_2)}\frac{d}{d\alpha_2}\left(\sin\alpha_2\frac{d\Phi_2(\alpha_2)}{d\alpha_2}\right) +$$
$$+\frac{\sin^2\alpha_2}{\Phi_3(\alpha_3)}\frac{d}{d\alpha_3}\left(\sin^2\alpha_3\frac{d\Phi_3(\alpha_3)}{d\alpha_3}\right) + ... + \frac{\sin^2\alpha_2...\sin^2\alpha_{p-3}}{\Phi_{p-2}(\alpha_{p-2})}\frac{d}{d\alpha_{p-2}}\left(\sin^2\alpha_{p-2}\frac{d\Phi_{p-2}(\alpha_{p-2})}{d\alpha_{p-2}}\right). \tag{3.19}$$

The left part of (3.19) is depended at angle α_1

(3.19) depended at angles $\alpha_2,...,\alpha_{p-2}$

to same constant C_3

$$-\frac{1}{\Phi_1(\alpha_1)}\frac{d^2\Phi_1(\alpha_1)}{d\alpha_1^2} = C_3 . \tag{3.20}$$

From the right part (3.19) there is

$$C_3 = C_2\sin^2\alpha_2...\sin^2\alpha_{p-2} + \frac{\sin\alpha_2}{\Phi_2(\alpha_2)}\frac{d}{d\alpha_2}\left(\sin\alpha_2\frac{d\Phi_2(\alpha_2)}{d\alpha_2}\right) +$$
$$+\frac{\sin^2\alpha_2}{\Phi_3(\alpha_3)}\frac{d}{d\alpha_3}\left(\sin^2\alpha_3\frac{d\Phi_3(\alpha_3)}{d\alpha_3}\right) + ... + \frac{\sin^2\alpha_2...\sin^2\alpha_{p-3}}{\Phi_{p-2}(\alpha_{p-2})}\frac{d}{d\alpha_{p-2}}\left(\sin^2\alpha_{p-2}\frac{d\Phi_{p-2}(\alpha_{p-2})}{d\alpha_{p-2}}\right). \tag{3.21}$$

The equation (3.21) can rewrite as follows

$$\frac{C_3}{\sin^2\alpha_2}-\frac{1}{\Phi_2(\alpha_2)\sin\alpha_2}\frac{d}{d\alpha_2}\left(\sin\alpha_2\frac{d\Phi_2(\alpha_2)}{d\alpha_2}\right)=C_2\sin^2\alpha_3...\sin^2\alpha_{p-2}+$$

$$+\frac{1}{\Phi_3(\alpha_3)}\frac{d}{d\alpha_3}\left(\sin^2\alpha_3\frac{d\Phi_3(\alpha_3)}{d\alpha_3}\right)+...+\frac{\sin^2\alpha_3...\sin^2\alpha_{p-3}}{\Phi_{p-2}(\alpha_{p-2})}\frac{d}{d\alpha_{p-2}}\left(\sin^2\alpha_{p-2}\frac{d\Phi_{p-2}(\alpha_{p-2})}{d\alpha_{p-2}}\right). \tag{3.22}$$

The left part of (3.22) is depended at angle α_2

(3.22) depended at angles $\alpha_3,...,\alpha_{p-2}$

to same constant C_4

$$\frac{C_3}{\sin^2\alpha_2}-\frac{1}{\Phi_2(\alpha_2)\sin\alpha_2}\frac{d}{d\alpha_2}\left(\sin\alpha_2\frac{d\Phi_2(\alpha_2)}{d\alpha_2}\right)=C_4 . \tag{3.23}$$

From the right part (3.22) there is

$$C_4=C_2\sin^2\alpha_3...\sin^2\alpha_{p-2}+$$

$$+\frac{1}{\Phi_3(\alpha_3)}\frac{d}{d\alpha_3}\left(\sin^2\alpha_3\frac{d\Phi_3(\alpha_3)}{d\alpha_3}\right)+...+\frac{\sin^2\alpha_3...\sin^2\alpha_{p-3}}{\Phi_{p-2}(\alpha_{p-2})}\frac{d}{d\alpha_{p-2}}\left(\sin^2\alpha_{p-2}\frac{d\Phi_{p-2}(\alpha_{p-2})}{d\alpha_{p-2}}\right). \tag{3.24}$$

The equation (3.24) can rewrite as follows

$$\frac{C_4}{\sin^2\alpha_3}-\frac{1}{\Phi_3(\alpha_3)\sin^2\alpha_3}\frac{d}{d\alpha_3}\left(\sin^2\alpha_3\frac{d\Phi_3(\alpha_3)}{d\alpha_3}\right)=C_2\sin^2\alpha_4...\sin^2\alpha_{p-2}+$$

$$+\frac{1}{\Phi_4(\alpha_4)}\frac{d}{d\alpha_4}\left(\sin^2\alpha_4\frac{d\Phi_4(\alpha_4)}{d\alpha_4}\right)+...+\frac{\sin^2\alpha_4...\sin^2\alpha_{p-3}}{\Phi_{p-2}(\alpha_{p-2})}\frac{d}{d\alpha_{p-2}}\left(\sin^2\alpha_{p-2}\frac{d\Phi_{p-2}(\alpha_{p-2})}{d\alpha_{p-2}}\right). \tag{3.25}$$

From the preceding it is clear that the process of separating equations for unknown angles can be continued to the end, i.e., until a complete system of equations for these unknowns is obtained. This, in principle, allows us to solve a system of equations for any finite value of p; for any finite dimension of the space, in principle, obtain a solution of the Schrödinger equation. For $p = 4$, this solution was obtained in the monograph (Zhizhin, 2018), and for p

= 5, in the monograph (Zhizhin, 2019). Let, for example, the dimension of the space be five ($p = 5$). Then the equation (3.17) takes the form

$$C_1 \sin\alpha_4^{\ 2} + \frac{1}{\Phi_4(\alpha_4)}\frac{d}{d\alpha_4}\left(\sin\alpha_4^{\ 2}\frac{d\Phi_4(\alpha_4)}{d\alpha_4}\right) = -\frac{1}{\Phi_1(\alpha_1)\sin\alpha_2^{\ 2}\sin^2_3}\frac{d^2\Phi_1(\alpha_1)}{d\alpha_1^{\ 2}} - \\ -\frac{1}{\Phi_2(\alpha_2)\sin\alpha_2\sin^2\alpha_3}\frac{d}{d\alpha_2}\left(\sin\alpha_2\frac{d\Phi_2(\alpha_2)}{d\alpha_2}\right) - \frac{1}{\Phi_3(\alpha_3)\sin^2\alpha_3}\frac{d}{d\alpha_3}\left(\sin^2\alpha_3\frac{d\Phi_3(\alpha_3)}{d\alpha_3}\right). \tag{3.26}$$

Since both parts of the equation (3.26) depend on different variables, they are equal to some constant. From left part equation (3.26) there is

$$C_1 \sin\alpha_4^{\ 2} + \frac{1}{\Phi_4(\alpha_4)}\frac{d}{d\alpha_4}\left(\sin\alpha_4^{\ 2}\frac{d\Phi_4(\alpha_4)}{d\alpha_4}\right) = C_2. \tag{3.27}$$

From right part of equation (3.26) there is

$$\frac{1}{\Phi_1(\alpha_1)}\frac{d^2\Phi_1(\alpha_1)}{d\alpha_1^{\ 2}} = C_2\sin^2\alpha_2\sin^2\alpha_3 + \frac{\sin\alpha_2}{\Phi_2(\alpha_2)}\frac{d}{d\alpha_2}\left(\sin\alpha_2\frac{d\Phi_2(\alpha_2)}{d\alpha_2}\right) + \\ + \frac{\sin^2\alpha_2}{\Phi_3(\alpha_3)}\frac{d}{d\alpha_3}\left(\sin^2\alpha_3\frac{d\Phi_3(\alpha_3)}{d\alpha_3}\right). \tag{3.28}$$

Since both parts of the equation (3.28) depend on different variables, they are equal to some constant. From left part equation (3.28) can obtain equation (3.20).

The equation (3.20) for any dimension p have decision

$$\Phi_1(\alpha_1) = Ae^{\pm i\sqrt{C_3}\alpha_1}. \tag{3.29}$$

Since

$$\Phi_1(0) = \Phi_1(2\pi),$$

then

$$A = Ae^{\pm i\sqrt{C_3}2\pi}, \tag{3.30}$$

i.e., $e^{\pm i\sqrt{C_3}2\pi} = 1$. Consequently, $\cos\left(2\pi\sqrt{C_3}\right) \pm i\sin\left(2\pi\sqrt{C_3}\right) = 1$.
It may be only if $\sqrt{C_3} = 0, \pm 1, \pm 2, \ldots$.
We got the first quantum number $m_1 = \sqrt{C_3} = 0, \pm 1, \pm 2, \ldots$

The constant A is defined from condition of normalization

$$\int_0^{2\pi} \Phi_1^*(\alpha_1)\Phi_1(\alpha_1)d\alpha_1 = 1 = A^2 \int_0^{2\pi} e^{-i\sqrt{C_3}\alpha_1} e^{i\sqrt{C_3}\alpha_1} d\alpha_1 = A^2 2\pi.$$

Consequently, $A = \frac{1}{\sqrt{2\pi}}$. Thus, the function $\Phi_1(\alpha_1)$ it is known for any p

$$\Phi_1(\alpha_1) = \frac{1}{\sqrt{2\pi}} e^{-im_1\alpha_1}, m_1 = \sqrt{C_3}. \tag{3.32}$$

From left part equation (3.28) there is

$$\frac{C_3}{\sin^2\alpha_2} - \frac{1}{\Phi_2(\alpha_2)\sin\alpha_2}\frac{d}{d\alpha_2}\left(\sin\alpha_2 \frac{d\Phi_2(\alpha_2)}{d\alpha_2}\right) = C_2\sin^2\alpha_3 + \\ + \frac{1}{\Phi_3(\alpha_3)}\frac{d}{d\alpha_3}\left(\sin^2\alpha_3 \frac{d\Phi_3(\alpha_3)}{d\alpha_3}\right). \tag{3.33}$$

Since both parts of the equation (3.33) depend on different variables, they are equal to some constant. From left part equation (3.33) we have equation (3.23). The equation (3.23) can rewrite with account (3.31) as follows

$$\Phi_2(\alpha_2)\left(C_4 - \frac{m_1^2}{\sin^2\alpha_2}\right) + \frac{1}{\sin\alpha_2}\frac{d}{d\alpha_2}\left(\sin\alpha_2 \frac{d\Phi_2(\alpha_2)}{d\alpha_2}\right) = 0. \tag{3.34}$$

The equation (3.34) is the adjoined Legendre equation. In case $C_4 = l(l+1)$ and we were l are integers, its solutions are the adjoined Legendre polynomials (Landau & Lifshitz, 1963)

$$P_l^{m_1}(\cos\alpha_2) = \frac{1}{2^l l!}\sin^{m_1}\alpha_2 \frac{d^{m_1+l}}{(d\cos\alpha_2)^{m_1+l}}(\cos^2\alpha_2 - 1)^l. \tag{3.35}$$

We can get second quantum number l – orbital quantum number. From right part equation (3.33) there is

$$C_2\sin^2\alpha_3 + \frac{1}{\Phi_3(\alpha_3)}\frac{d}{d\alpha_3}\left(\sin^2\alpha_3\frac{d\Phi_3(\alpha_3)}{d\alpha_3}\right) = \mathrm{C}_4. \tag{3.36}$$

One rewrite the equation (3.36) in the form

$$\left(C_2 - l\frac{l+1}{\sin^2\alpha_3}\right)\Phi_3(\alpha_3) + \frac{1}{\sin^2\alpha_3}\frac{d}{d\alpha_3}\left(\sin^2\alpha_3\frac{d\Phi_3(\alpha_3)}{d\alpha_3}\right) = 0. \tag{3.37}$$

Since $\sin^2\alpha_3$ is a positive function varying from 0 to 1, then we introduce the average value of this function τ in the range α_3 from 0 to 2π

$$\tau = \frac{1}{2\pi}\int_0^{2\pi}\sin^2\alpha_3 d\alpha_3 = \frac{1}{2}.$$

Then the equation has form

$$C_2 - 2l(l+1) = -\frac{1}{\Phi_3(\alpha_3)}\frac{d^2\Phi_3(\alpha_3)}{d\alpha_3^2}. \tag{3.38}$$

The decision of this equation is

$$\Phi_3(\alpha_3) = Be^{\pm i\sqrt{C_2-2l(l+1)}\alpha_3}. \tag{3.39}$$

Since $\Phi_3(0) = \Phi_3(2\pi)$, $B = Be^{\pm i\sqrt{C_2-2l(l+1)}2\pi}$,

i.e., $e^{\pm i\sqrt{C_2-2l(l+1)}2\pi} = 1$. Consequently,

$$\cos\left(2\pi\sqrt{C_2-2l(l+1)}\right) \pm i\sin\left(2\pi\sqrt{C_2-2l(l+1)}\right) = 1.$$

It may be only if $\sqrt{C_2-2l(l+1)} = 0, \pm 1, \pm 2, \ldots$

We can get the third quantum number

$$m_2 = \sqrt{C_2-2l(l+1)} = 0, \pm 1, \pm 2, \ldots. \tag{3.40}$$

The constant B is defined from condition of normalization

$$\int_0^{2\pi} \Phi_3^*(\alpha_3)\Phi_3(\alpha_3)d\alpha_3 = 1 = B^2\int_0^{2\pi} e^{-i\sqrt{C_2-2l(l+1)}\alpha_3}e^{i\sqrt{C_2-2l(l+1)}\alpha_3}d\alpha_3 = B^2 2\pi.$$

Consequently, $B = \frac{1}{\sqrt{2\pi}}$. Thus, the function $\Phi_3(\alpha_3)$ it is known

$$\Phi_3(\alpha_3) = \frac{1}{\sqrt{2\pi}}e^{-im_2\alpha_3}, m_2 = \sqrt{C_2-2l(l+1)}. \tag{3.41}$$

We now turn to equation (3.27) with respect to the angle α_4
The equation (3.37) can rewrite as follows

$$\left(C_1 - 2l\frac{l+1}{\sin^2\alpha_4}\right)\Phi_4(\alpha_4) + \frac{1}{\sin^2\alpha_4}\frac{d}{d\alpha_4}\left(\sin^2\alpha_4\frac{d\Phi_4(\alpha_4)}{d\alpha_4}\right) = 0. \tag{3.44}$$

Performing the transformations in equation (3.44) for the angle α_4 the same as in equation (3.37) for the angle α_3
(3.44) to the form

$$C_1 - 4l(l+1) = -\frac{1}{\Phi_4(\alpha_4)}\frac{d^2\Phi_4(\alpha_4)}{d\alpha_4^{\ 2}} . \tag{3.45}$$

The equation (3.45), so equation (3.38), have decision

$$\Phi_4(\alpha_4) = \frac{1}{\sqrt{2\pi}}e^{-im_3\alpha_4}, \quad m_3 = \sqrt{C_1 - 4l(l+1)} = 0; \pm 1, \pm 2, \ldots$$

quantum number. (3.46)

We now consider the equation for the radius (3.15). For $p = 5$ and $C_1 = m_3^2 + 4l(l)$ this equation has the form

$$\frac{1}{Rr^2}\frac{d}{dr}\left(r^4\frac{dR}{dr}\right) + 2r^2\left(\frac{1}{r} + E\right) = m_3^2 + 4l(l+1).$$

Performing differentiation in this equation, can obtain

$$r^2\frac{d^2R}{dr^2} + 4r\frac{dR}{dr} + R\left(2r + 2Er^2 - 4l(l+1) - m_3^2\right) = 0. \tag{3.47}$$

We seek the solution of the equation (3.47) in the form of a series in powers of r

$$R(r) = e^{-\mu r}r^l\sum_j b_j r^j . \tag{3.48}$$

In (3.48) $\mu = \sqrt{-2E}, E < 0, b_j$ are numerical coefficients. One substitute (3.48) into (3.47)

$$\sum_j b_j r^{l+j}\left((l+j)^2+2(l+j)-4l(l+1)-m_3^2\right)=\sum_j b_j r^{l+j+1}\left(2\mu(l+j+2)-2\right). \tag{3.49}$$

From equation (3.49) follows the recurrence formula for the coefficients

$$b_{j+1}=\frac{2\mu(l+j+2)-2}{(l+j)^2+2(l+j)-4l(l+1)-m_3^2}b_j. \tag{3.50}$$

In view of the need to obtain a limited physical meaningful solution, the values of the coefficients, starting from a certain value of j must vanish. This is possible if the numerator on the right-hand side of (3.50) is zero

$$\mu(l+j+2)-1=0 \tag{3.51}$$

From equation (3.51) follows that

$$\mu=\sqrt{-E}=\frac{1}{l+j+2}. \tag{3.52}$$

We introduce the notation $n = l + j + 1$ is principal quantum number. From (3.52) equation can obtain

$$E=-\frac{1}{2(n+1)^2}. \tag{3.53}$$

The radial function is determined by the equality (3.48) of setting different values j. For example, with $j = 0$

$$R(r)=\mathrm{e}^{-\mu r}r^l b_0, b_0=\frac{1}{\int\limits_0^\infty e^{-\mu r}r^l dr}.$$

With $j = 1$

$$R(r) = e^{-\mu r} r^l b_0 \left(1 - r \frac{2\mu(l+2) - 2}{3l^2 + 2l + m_3^2} \right),$$

$$b_0 = \frac{1}{\int_0^{\infty} e^{-\mu r} r^l \left(1 + \frac{2 - 2\mu(l+2)}{3l^2 + 2l + m_3^2} \right) dr}.$$

Thus, one can obtain various approximations for the function $R(r)$.

3.4.3. Quantum Numbers of Solutions of the Schrödinger Equation in a Space of Higher Dimension

The study showed that the motion of an electron around a nucleus in a space of higher dimension leads to the appearance of quantum numbers with a total number equal to the dimension of space. On a level with the principal quantum number, characterizing the average distance of the electron from the nucleus, and the orbital quantum number, characterizing the shape of the electron orbitals, a series of magnetic quantum numbers appear in a space of higher dimension. Its correspondent the possible values of the projections of the angular momentum of an electron on the coordinate axes of space. The number of these magnetic quantum numbers is equal to the dimension of the space minus two. Taking into account the principal quartic number and orbital quantum number, the total number of quantum numbers characterizing the motion of an electron around the nucleus is equal to the dimension of space. Thus, the total number of quantum numbers of an electron with allowance for the spin quantum number characterizing the internal rotation of an electron is equal to the dimension of space plus one. In this case, the electron energy in the first approximation, as well as in the model of the Bohr atom, is inversely proportional to the square of the principal quantum number (equation (3.51)). For example, in three-dimensional space the atom has besides a spin quantum number yet of three more quantum numbers. The main quantum number n, which determines the average distance of the electron from the nucleus. Orbital quantum number l, characterizing the momentum of an electron. The magnetic quantum number m_l, characterizing the possible value of the projection of the angular momentum of the electron on the axis z in a magnetic field. The main quantum number can take the values 1, 2, 3, ... The orbital quantum number can take the values 0, 1, 2, ...

Magnetic quantum numbers can take the values 0, ±1, ±2, ±3,... .

The values of the orbital quantum number *l* correspond to form of electron orbital. Taking into account the large dimensionality of the neighborhood of the nucleus in the atom, one should expect an increase in the number of quantum cells in orbitals *p, d, f* while maintaining their shape. For example, not 6 but 8 electrons (for each quantum cell with a corresponding pair of electrons per axis) can be on the orbital *p*. When analyzing chemical elements and molecules formed by them, it was shown (Zhizhin, 2018) that many elements exhibit a valence in chemical reactions much higher than the valence determined by their place in the periodic table (group number). To explain this, can resort to assumptions about various possible mechanisms of electron interaction: atomic and molecular hybridization, repulsion of divided and unpaired electron pairs, and the involvement of electrons inside the electron shells in the chemical interactions. In addition, physical experiments at high pressures reveal a sharp change in the structure of compounds that are also not explainable from the point of view of standard valence.

On the other hand, the analysis of real forms of molecules shows that their dimensionality, determined by the shape of the corresponding convex body, whose vertices are atoms, is often higher than three.

Conclusion

The research has shown that the laws of incidence, acting in the geometry of polytopes of higher dimension, contradict the axioms of the connection of the *n*-dimensional geometry of Euclidean. The axiom of parallelism of Euclidean geometry, based on the concept of infinity of space, is also inapplicable in the geometry of higher-dimensional polytopes due to the limited spaces of higher-dimensional polytopes. Therefore, the geometry of polytopes of the highest dimension should be recognized as non-Euclidean. Moreover, this property is not associated with the curvature of space, as is the case in hyperbolic and elliptic geometries, but is associated with a special structure characteristic of all polytopes of the highest dimension.

The study showed that Riemann's position on the limitedness of space, combined with ideas about its higher dimension, taking into account modern knowledge about the structure of higher-dimensional space, lead to specific properties of the inertial motion of material points in these spaces. If a material point belongs to the set of internal points of space of the highest dimension, then in its inertial motion it can remain in this set, successively reflecting from the edges of the boundary complex. If a material point in its inertial motion has sufficient energy, then it crosses the nearest flat face of the boundary complex, falls into the corresponding three-dimensional space of the boundary complex, and then, crossing other flat faces of the boundary complex, falls into the faces of different dimensions of the boundary complex, successively disappearing from these spaces of different dimensions and arising in other spaces of the boundary complex of different dimensions. At the same time, the boundary complex retains its integrity.

It is proved that the Schrödinger equation has a solution in a space of higher dimension. Consequently, the discrete nature of matter exists in a space of higher dimension. The incorrectness of authors in previous studies of the solution of the Schrodinger equation in space of higher dimension was that the potential energy of electron was written in form different at law of Coulomb. The study showed that the motion of an electron around a nucleus in a space of higher dimension leads to the appearance of quantum numbers with a total number equal to the dimension of space. When analyzing chemical elements and molecules formed by them, it was shown (Zhizhin, 2018) that many elements exhibit a valence in chemical reactions much higher than the valence determined by their place in the periodic table (group number). This may be due to the higher dimension of space in the vicinity of atomic nuclei and, in accordance with the obtained solution of the Schrödinger equation, with an increase in the number of quantum numbers. Undoubtedly, the possible loss of the number of quantum cells in orbitals *p, d, f* should not lead to a change in the number of electrons in the atomic structure in neutral atoms. The number of electrons under these conditions is of course equal to the charge of the nucleus. But the possible increase in the number of vacant quantum cells contributes to an increase in the chemical activity of atoms, allows the creation of complex chemical compounds. This can see, for example, in biology.

Since almost all chemical compounds of animate and inanimate nature on planet Earth have the highest dimension, the geometry of all these chemical compounds should be considered non-Euclidean.

Chapter 4

Polytopes of the Highest Dimension of Inert Substances

Abstract

The structures of inanimate (inert) natural substances (including precious stones) are considered. It is shown that all these structures have the highest dimension. The greater the number of different chemical elements included in the unit cell of the structure, the higher its dimension. It is emphasized that the formation of complex structures of inanimate (inert) substances is the result of self-organization processes taking place in them. It was found that the structure of the garnet cannot be explained even using the concept of the highest dimensionality of the garnet space. In this regard, the term "inert," introduced by V.I. Vernadsky to denote a non-living natural substance, should be considered non-reflective of the essence of these substances.

Keywords: higher-dimensional space, Euclidean geometry, non-Euclidean geometry, polytope, polytopic prismahedron, incidence

Introduction

After studying the geometry of higher-dimensional polytopes and proving that this geometry is non-Euclidean, which differs from non-Euclidean hyperbolic and elliptic geometries, one should proceed to the analysis of polytopes corresponding (according to Vernadsky's definition, Vernadsky, 1925, 1977) to molecules of inert and living matter. This chapter will consider the geometry of molecules of inert natural matter adamantane, wurtzite, fluorite, etc. Consideration of the geometry of compounds begins with natural chemical compounds of atoms of one type, then binary compounds, i.e., compounds in which atoms of two types participate, and the chapter ends with a consideration of natural compounds in which atoms of several types participate. In all cases, the highest dimension of these natural formations and the fulfillment of the Euler-Poincaré equation for them with the corresponding dimension (Poincaré, 1895) is proved.

4.1. The Dimension of Adamantane Molecules and Methods of Molecules Connecting with Each Other

As a chemical compound, adamantine was discovered in 1933 (Landa, Machacek, 1933). Adamantane molecule consists of 10 carbon atoms (atoms one types), repeating disposition of carbon atoms in the diamond crystal lattice, and 16 hydrogen atoms connected to carbon atoms by their valence links unsaturated with carbon atoms. Adamantane discovery served as an impulse for the development of organic polyhedranes chemistry. Derivatives of adamantane (e.g., amantadine, memantine, rimantadine, tromantadine) have found practical application in medicine as pharmaceuticals of different biological activity and purpose (as antiviral, antispasmodic, anti-Parkinson drugs, etc.). All these drugs have the same structural group of carbon atoms which is peculiar to adamantane, only structural groups connected to carbon atoms change. Among inorganic and organ-elemental compounds, there are many structural analogs of adamantane molecule, such as phosphorus oxide, urotropine and others. In 2005, a silicon analogue of adamantane was synthesized (Fischer, Baumgartner, Marschner, 2005) (as known, silicon amounts to 27.6%). In scientific literature (Bauschlicher, et al., 2007; Dahl, Lie, Carlson, 2003), for example, – adamantane is usually depicted as it is shown in Figure 4.1.

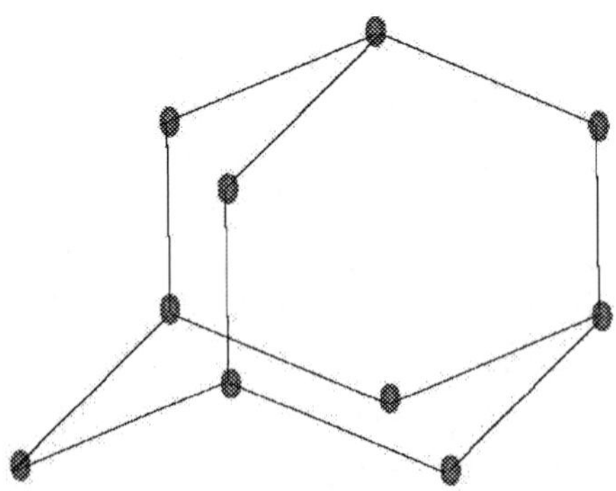

Figure 4.1. Schema of the adamantane molecule.

The adamantane structure is a common one. As a rule, hydrogen atoms are not depicted. Speaking hereinafter about adamantane molecule, we'll often keep in mind exactly 10 carbon atoms of adamantane molecule, although, strictly speaking, it is only a part of it. However, Figure 4.1 gives us little information and does not reflect the main features of spatial arrangement of atoms.

4.1.1. The Dimension of the Adamantine Molecule

4.1.1.1. Theorem 4.1 (Zhizhin, 2014 a, b)

The adamantane molecule is a convex polytope in the 4D space.

4.1.1.2. Proof

Let's construct an adamantane cell taking into consideration that 6 from 10 carbon atoms of adamantane are located in the centers of flat faces of the cube $\alpha_2, \alpha_6, \alpha_3, \alpha_4, \alpha_8, \alpha_9$

the cube $\alpha_1, \alpha_5, \alpha_7, \alpha_{10}$

cube flat faces and relevant to the cube $\beta_1, \ldots \beta_8$ (Figure 4.2) vertices common to these faces.

On Figure 4.2 except for the edges of the cube (thin dashed line) are emerging in the construction of cell molecules lines of different kinds. Solid thick lines represent valence bonds between the carbon atoms; at three such links from each carbon $\alpha_1, \alpha_5, \alpha_7, \alpha_{10}$ located inside $\beta_1, \ldots, \beta_8$ cube. The fourth connection from each of these carbon atoms is directed toward the vertices of the cube in which in the case of diamond carbon atoms are also located. These links are indicated in Figure 4.2 by thick dotted lines to mark the location of these connections is adamantine molecule. Solid thin lines delineate the regular tetrahedron inscribed in a cube, its edges are the diagonals of the faces of the cube. Thick bar dotted lines delineate the regular octahedron, passing through the atoms located in the cube centers $\alpha_2, \alpha_6, \alpha_3, \alpha_4, \alpha_8, \alpha_9$. Thin dash-dotted lines delineate the regular tetrahedron whose vertices coincide with the carbon atoms in a cube $\alpha_1, \alpha_5, \alpha_7, \alpha_{10}$.

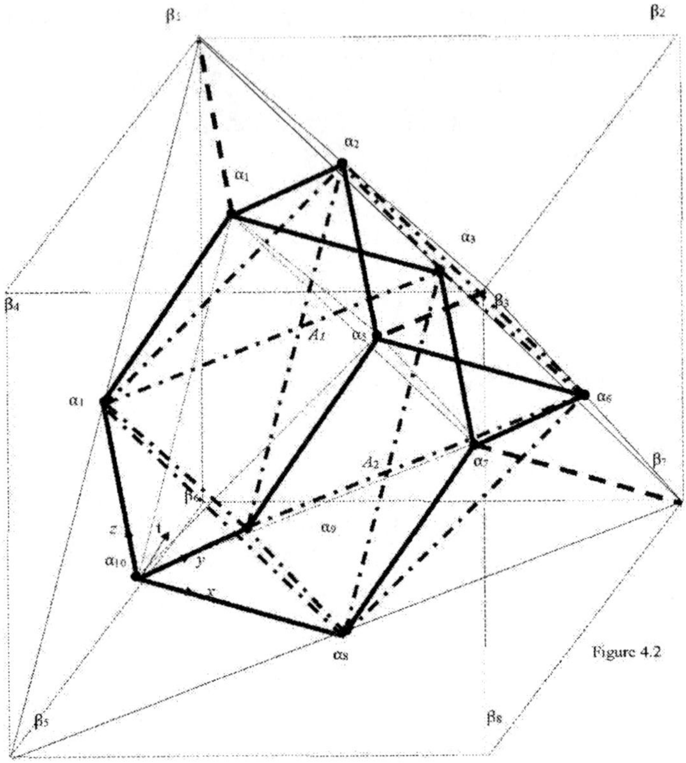

Figure 4.2. The structure of molecule adamantine.

By construction, the formed segments connecting the vertices of adamantane split into 10 families of parallel segments, each family including three parallel segments: 1) $\alpha_1\alpha_2, \alpha_7\alpha_6, \alpha_9\alpha_{10}$; 2) $\alpha_1\alpha_3, \alpha_8\alpha_{10}, \alpha_5\alpha_6$; 3) $\alpha_3\alpha_2, \alpha_7\alpha_5, \alpha_8\alpha_9$; 4) $\alpha_1\alpha_4, \alpha_7\alpha_8, \alpha_5\alpha_9$; 5) $\alpha_4\alpha_2, \alpha_8\alpha_6, \alpha_{10}\alpha_5$; 6) $\alpha_9\alpha_2, \alpha_3\alpha_8, \alpha_{10}\alpha_1$; 7) $\alpha_4\alpha_3, \alpha_7\alpha_{10}, \alpha_6\alpha_9$; 8) $\alpha_2\alpha_5, \alpha_7\alpha_3, \alpha_4\alpha_{10}$; 9) $\alpha_1\alpha_5, \alpha_3\alpha_6, \alpha_4\alpha_9$; 10) $\alpha_2\alpha_6, \alpha_7\alpha_1, \alpha_4\alpha_8$

segments (each of them is an edge of a polyhedron) is equal to 30. The length of segments is determined from the length of cube edges. Let's assume that the length of cube edge is equal to 1 (one should enter a scale factor to receive the specific dimension of a bond length). Then regular tetrahedrons with the bases on the faces of octahedron and the vertices coinciding with cube vertices (for example tetrahedron $\beta_1\alpha_2\alpha_3\alpha_4$

length $a = 1/\sqrt{2}$ and the radius of circles described around the tetrahedrons is $b=\sqrt{3}/4$ (the points $\alpha_2, \alpha_6, \alpha_3, \alpha_4, \alpha_8, \alpha_9$

Therefore, segments 2), 5), 6), 7), 9), 10) have the length *a*, while the segments 1), 3), 4), 8) have the length *b*. Thus, the two-dimensional geometric elements involved in adamantane have as sides the segments with lengths *a*, and *b*. One can define (Figure 4.2) that a set of two-dimensional faces belonging to adamantane form regular triangles with sides *a*, an isosceles triangle with the base *a*, and two sides *b*, squares with sides *a*, and rectangles with sides *a*, and *b*. Among the regular triangles, there are 4 triangles located at the outer edge of adamantine ($\alpha_2\alpha_6\alpha_3, \alpha_4\alpha_2\alpha_9, \alpha_4\alpha_3\alpha_8, \alpha_9\alpha_6\alpha_8$

adamantine ($\alpha_1\alpha_5\alpha_{10}, \alpha_1\alpha_7\alpha_{10}, \alpha_1\alpha_7\alpha_5, \alpha_9\alpha_6\alpha_2$, $\alpha_2\alpha_4\alpha_3, \alpha_8\alpha_3\alpha_6, \alpha_4\alpha_9\alpha_8, \alpha_5\alpha_{10}\alpha_7$

Among the irregular triangles, there are 12 triangles located at the outer edge of adamantine ($\alpha_2\alpha_1\alpha_3, \alpha_1\alpha_2\alpha_4, \alpha_4\alpha_3\alpha_1, \alpha_9\alpha_2\alpha_5$, $\alpha_2\alpha_6\alpha_5, \alpha_3\alpha_7\alpha_8, \alpha_7\alpha_3\alpha_6, \alpha_9\alpha_4\alpha_{10}$

$\alpha_4\alpha_{10}\alpha_8, \alpha_5\alpha_6\alpha_9, \alpha_7\alpha_6\alpha_8, \alpha_9\alpha_{10}\alpha_8$) and 6 triangles located in the inner part of adamantane ($\alpha_1\alpha_4\alpha_{10}, \alpha_5\alpha_2\alpha_1, \alpha_5\alpha_6\alpha_7, \alpha_1\alpha_7\alpha_3, \alpha_{10}\alpha_5\alpha_9, \alpha_7\alpha_8\alpha_{10}$). Thus, there are in total 30 triangles in adamantane. In the inner part of adamantane there are three squares with side a as three sections of the octahedron, ($\alpha_2\alpha_6\alpha_8\alpha_4, \alpha_2\alpha_9\alpha_3\alpha_8, \alpha_9\alpha_6\alpha_3\alpha_4$

sides a and b ($\alpha_1\alpha_3\alpha_8\alpha_{10}, \alpha_2\alpha_9\alpha_1\alpha_{10}, \alpha_9\alpha_1\alpha_5\alpha_4$, $\alpha_1\alpha_7\alpha_8\alpha_4, \alpha_1\alpha_5\alpha_3\alpha_6, \alpha_1\alpha_6\alpha_7\alpha_2$,

$\alpha_2\alpha_5\alpha_{10}\alpha_4, \alpha_2\alpha_7\alpha_3\alpha_5, \alpha_7\alpha_{10}\alpha_3\alpha_4$, $\alpha_5\alpha_6\alpha_8\alpha_{10}, \alpha_5\alpha_9\alpha_7\alpha_8, \alpha_9\alpha_6\alpha_7\alpha_{10}$

These parallelograms are rectangles, as one can prove that the planes of irregular triangles, resting upon the sides of specified squares, are perpendicular to the planes of these squares. Indeed, let's cut up the adamantane by a plane passing, for example (see Figure 4.2), through the top α_2 and the edges $\alpha_1\alpha_2, \alpha_2\alpha_5$

tetrahedron built on its edges, these edges lie in the same plane). This plane cuts up irregular triangles $\alpha_1\alpha_3\alpha_4, \alpha_5\alpha_9\alpha_6$

$\alpha_2\alpha_3\alpha_4, \alpha_8\alpha_9\alpha_6$ at their heights, passing through the middle of the edges $\alpha_3\alpha_4, \alpha_9\alpha_6$ (respectively the points A_1, A_2 α_8.

Intersection plane is presented in Figure 4.3.

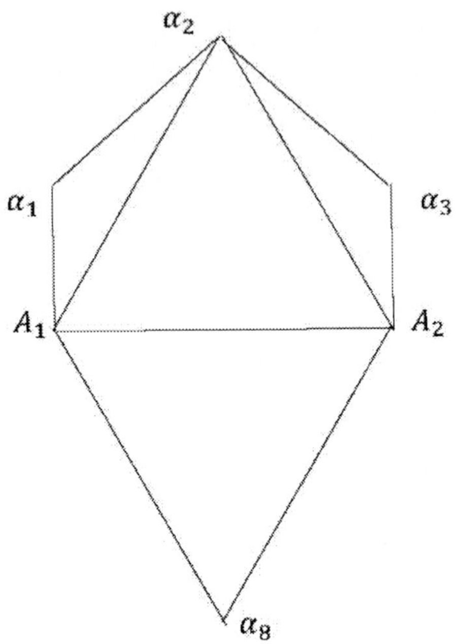

Figure 4.3. Section of adamantine.

Let's prove that the segments $A_1\alpha_1$ and $\alpha_3 A_2$

line A_1A_2

perpendicular to the plane of the square $\alpha_6\alpha_3\alpha_9\alpha_4$

triangle $\alpha_2\alpha_3A_2$; in it $A_1A_2 = a, \alpha_2A_2 = b = a\sqrt{6}/4, A_2\alpha_3 = a/(2\sqrt{2})$.

$\cos\angle\alpha_2A_2\alpha_3 = \sqrt{2/3}, \sin\angle\alpha_2A_2\alpha_3 = 1/\sqrt{3}$. From the triangle $A_1A_2\alpha_2$ can see that $\cos\angle\alpha_2A_2A_1 = \sqrt{1/3}, \sin\angle\alpha_2A_2A_1 = \sqrt{2/3}$. Consequently, $\cos(\angle\alpha_2A_2\alpha_3 + \angle\alpha_2A_2A_1) = 0$. Then, $\alpha_3A_2 \perp A_1A_2$

This also implies that $\alpha_1\alpha_3 \perp \alpha_3\alpha_6, \alpha_5\alpha_6 \perp \alpha_3\alpha_6$

parallelogram $\alpha_3\alpha_6\alpha_1\alpha_5$ is a rectangle. One can also prove that the remaining parallelograms are also rectangles. Thus, the number of squares and rectangles is 15 and the total number of geometric elements of dimension 2 consisting of adamantane is 45.

These 2*D* geometric elements form in adamantane 25 of 3*D* polyhedron (Figure 4.2):

5 tetrahedrons ($\alpha_3\alpha_4\alpha_1\alpha_2, \alpha_{10}\alpha_7\alpha_1\alpha_5, \alpha_4\alpha_9\alpha_{10}\alpha_8, \alpha_3\alpha_6\alpha_7\alpha_8, \alpha_2\alpha_6\alpha_9\alpha_5$

prisms ($\alpha_3\alpha_2\alpha_1\alpha_8\alpha_9\alpha_{10}, \alpha_6\alpha_2\alpha_1\alpha_7\alpha_9\alpha_{10}, \alpha_3\alpha_2\alpha_5\alpha_8\alpha_9\alpha_7, \alpha_3\alpha_5\alpha_1\alpha_8\alpha_6\alpha_{10},$
$\alpha_5\alpha_2\alpha_6\alpha_8\alpha_4\alpha_{10}, \alpha_5\alpha_4\alpha_1\alpha_8\alpha_9\alpha_7$),

14 pyramids

($\alpha_4\alpha_2\alpha_1\alpha_9\alpha_{11}, \alpha_5\alpha_2\alpha_1\alpha_9\alpha_{10}, \alpha_4\alpha_3\alpha_1\alpha_8\alpha_{10}, \alpha_7\alpha_3\alpha_1\alpha_8\alpha_{10}, \alpha_4\alpha_2\alpha_3\alpha_9\alpha_8, \alpha_3\alpha_2\alpha_6\alpha_9\alpha_8, \alpha_3\alpha_2\alpha_1\alpha_5\alpha_6,$
$\alpha_5\alpha_3\alpha_1\alpha_6\alpha_7, \alpha_4\alpha_3\alpha_6\alpha_9\alpha_8, \alpha_4\alpha_2\alpha_3\alpha_9\alpha_6, \alpha_4\alpha_3\alpha_1\alpha_7\alpha_{10}, \alpha_4\alpha_3\alpha_7\alpha_{10}\alpha_8, \alpha_4\alpha_5\alpha_1\alpha_9\alpha_{10}, \alpha_4\alpha_2\alpha_1\alpha_5\alpha_9$),

When calculating *3D* figures octahedrons as the figures consisting of two pyramids were not considered, because square sections of octahedron are involved in the formation of other *3D* figures. Let's now calculate Euler-Poincaré formula for the polytope *P* of dimension *n* (Poincaré, 1895)

$$\sum_{i=0}^{n-1}(-1)^i f_i(P) = 1 + (-1)^{n-1}. \tag{4.1}$$

As previously was defined in this case we have

$$f_0(P) = 10, f_1(P) = 30, f_2(P) = 45, f_3(P) = 25.$$

There are no elements of dimension greater than 3 inside adamantane. Substituting the values obtained for the number of faces of different dimension in Euler-Poincare's formula, for $n = 4$, we obtain $10 - 30 + 45 - 25 = 0$, i.e., Euler-Poincare's formula for adamantane is true at $n = 4$. This proves statement of theorem 4.1.

Adamantane is irregular convex polytope of dimension 4. 6 rays emanate from each vertex of this polytope as in the 16-Cell convex regular *4D*-polytope, which consists of 16 regular tetrahedrons arranged on the octahedron base (Coxeter, 1963; Zhizhin, 2014 b). All two-dimensional faces of adamantane are simultaneously the faces of two or more three-dimensional figures, which indicates the closeness of adamantane as a polytope. The existence of the outer three-dimensional adamantane boundary consisting of two-dimensional faces doesn't contradict to adamantane four-dimensionality if we take into account the inner structure. Just as the above-mentioned four-dimensional 4 – cross-polytope can be considered as a figure consisting of two three-dimensional hexagonal pyramids applied to each other by their bases (Zhizhin, 2014 b). Only drawing inside this figure six edges which form two regular triangles makes this figure a four-dimensional polytope consisting of 16 tetrahedrons. The outer boundary of adamantane consisting of two-dimensional faces of the polytope is the projection of the polytope on the three-dimensional space, just as the outer boundary of any closed polytope on a two-dimensional plane is a closed circuit composed of one-dimensional segments.

4.1.2. Connection Types of Adamantane Molecules

In view of the established geometric properties of adamantane molecules they can contact each other in three ways: 1) at the vertices located in the centers of cube faces; 2) at the irregular parallel triangles; 3) at the broken hexagonal contours formed by a regular triangle and irregular triangles surrounding it and forming dihedral angles with the regular triangle.

Let's consider the first way of adamantane molecules joining.

It leads to the standard translational model of a diamond with a cube cell. At that, the coordinates of adamantane vertices are calculated as integers. Indeed, if we'll take the edge length of a unit cell equal to 4, the coordinates x, y, z of adamantane vertices in the initial position (let's denote it by index 0, $\alpha_{i,0}(x_{i,0}, y_{i,0}, z_{i,0})$

$$\alpha_{1,0}(1,1,3), \alpha_{2,0}(2,2,4), \alpha_{3,0}(2,0,2), \alpha_{4,0}(0,2,2), \alpha_{5,0}(3,3,3),$$
$$\alpha_{6,0}(4,2,2), \alpha_{7,0}(3,1,1), \alpha_{8,0}(2,2,0), \alpha_{9,0}(2,4,2), \alpha_{10,0}(1,3,1).$$

Then, at translation of the cube by $k(k_x, k_y, k_z)$ *x, y, z*, the coordinates of adamantane vertices are calculated accordingly

$$\alpha_{i,k}(x_{i,k}, y_{i,k}, z_{i,k}) = \alpha_{i,k}(x_{i,0} + 4k_x, y_{i,0} + 4k_y, z_{i,0} + 4k_z), i = 1, 2, ..., 10;$$

k_x, k_y, k_z being integers (positive and negative).

It is essential that to determine adamantane coordinates in integers is not necessary to use the theory of numbers, as was done in the work of Balaban (Balaban, 2013).

Studying a set of cubic cells including adamantane molecules, one can establish the existence of scaling process in the diamond, i.e., formation of large-scale geometric configurations from the same figures of a smaller scale. For the first time, the process of scaling was discovered in the phase transitions of the second order (Landau, 1937) and on the grid of giperrombohedron vertices in quasi-crystals (Zhizhin, 2013; 2014 b). Figure 4.4 shows the result of octahedron receiving on the basis of 8 cubes, each of them containing an octahedron 8 times smaller.

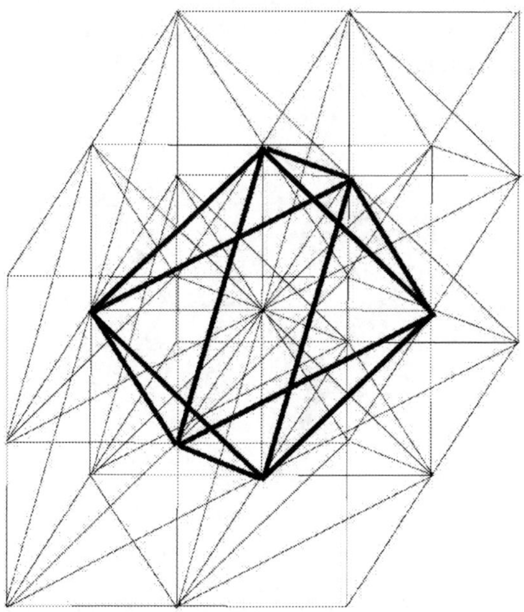

Figure 4.4. Scaling in diamond.

It explains the existence of diamond crystals of macroscopic dimensions with the same form as a microscopic unit cell of the diamond. The increase of scale at that occurs in a discrete manner. The scale of diamond crystal increases $n^3, (n = 1, 2, ...)$

Let's consider the second way of adamantane molecules joining.

Due to the fact established by the theorem, the plane of irregular triangle is perpendicular to the square section of octahedron. Therefore, adamantane molecules can be applied to each other in irregular triangles in two mutually perpendicular directions forming an infinite layer of adamantane polyhedrons. One layer may lie on the other contacting by free vertices of octahedrons. At the contact of adamantane polyhedrons by irregular triangles a dihedral angle forms between these triangles. Since the irregular triangles are perpendicular to the square section of octahedron, this angle is equal to the dihedral angle between irregular triangles in the adamantane polyhedron itself. Therefore, another adamantane polyhedron can be tightly nested in the space between the three adamantine polyhedrons contacting by irregular triangles. Thus, a completely filled infinite layer forms with a fundamental domain consisting of an adamantane polyhedron and a regular tetrahedron attached to it. These layers, contacting with each other, completely fill the space. Top views from all layers are shown in Figure 4.5 (thick lines in Figure 4.5 designate valence bonds).

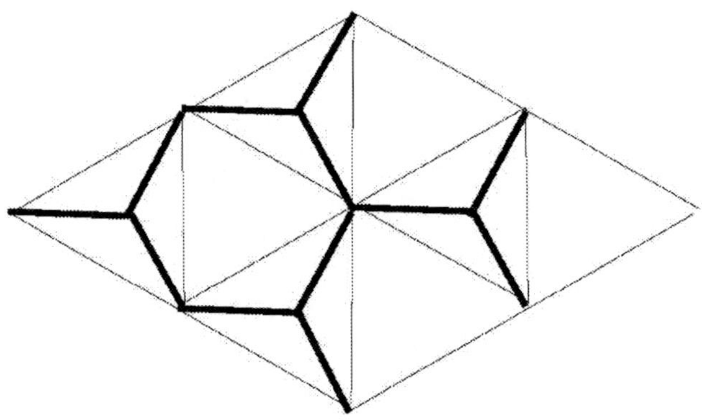

Figure 4.5. Adamantane connections in an oblique lattice.

One can see that the condition of carbon atoms four-valence in all the vertices is fulfilled, as the fourth valence bond in carbon atoms is directed from layer to layer (vertically). It is obvious that the distribution of carbon atoms according to Figure 4.5 is described by oblique coordinate system. The density of this atom`s arrangement must be higher than the density of a diamond with cubic fundamental domain.

Let's consider the third way of adamantane molecules joining.

The adamantane polyhedrons can contact each other not only by vertices and irregular triangles but also by broken hexagonal spatial contours whose edges correspond to the valence bonds. The boundary of each adamantane consists of 4 such contours. Combining two adamantanes in such a contour, we'll receive a diamantane. Connecting to it one adamantane more, cut along such outline, we'll receive a triamantane, and so on. Based on this formation mechanism, a general formula for compounds called diamondoids is

$$C_nH_{n+6},(n=4i+6,i=1,2,...)$$

All diamondoids indicated in (Bauschlicher, et al., 2007; Dahl, Liu, Carlson, 2003) submit to this formula.

These diamondoids are shells of the above-mentioned hexagonal contours. It's essential that these shells can connect with the structures of carbon atoms derived by the second method. At this even more complex and diverse compounds form.

Adamantane molecules are geometrically figures bounded by broken hexagons, i.e., each hexagon is formed by atoms which do not lie in one plane. Carbon atoms may form a structure in which they form pentagons. Such structures were called D_5 diamonds, they were investigated in works (Ashrafi, et al., 2013; Nagy, Diudea, 2013).

Adamantane is irregular convex polytope of dimension 4. From each vertex of this polytope outgoing 6 edges as in the 16-cell convex regular *4D*-polytope (Grunbaum, 1967; Zhizhin, 2014 b). All two-dimensional faces of adamantane are simultaneously the faces of two or more three-dimensional figures, which indicates the closeness of adamantane as a polytope. The existence of the outer three-dimensional adamantane boundary consisting of two-dimensional faces doesn't contradict to adamantane four-dimensionality if we take into account the inner structure. Just as the above-mentioned four-dimensional 4 – cross-polytope can be considered as a figure consisting of two three-dimensional hexagonal pyramids applied to each other by their bases (Zhizhin, 2014 a, b). Only drawing inside this figure six edges which form two regular triangles makes this figure a four-dimensional polytope consisting of 16 tetrahedrons. The outer boundary of adamantane consisting of two-dimensional faces of the polytope is the projection of the polytope on the three-dimensional space, just as the outer boundary of any closed polytope on a two-dimensional plane is a closed circuit composed of one-dimensional segments.

4.2. The Structure of Binary Natural Compounds

The various binary chemical compounds have a limited number of typical structures (Fersman, 1937). Since such structures form chlorides, bromides and iodides of alkali metals, we will refer to these structures rock salt structures. Silicon, germanium, tin and other elements lead also form a molecule of adamantane or a molecule with dimension 4 or more than four. For example, in crystalline beryllium fluoride, the linearity of the combination of beryllium atoms with fluorine atoms and the tetrahedral coordination of beryllium atoms with one another surprisingly are combined. In this case, a structure is formed topologically equivalent to the adamantane molecule, which includes 10 carbon atoms (see Figure 4.2). The difference from the adamantane molecule is the linear arrangement between the beryllium atoms of the fluorine atoms (Figure 4.6). Since the adamantane molecule has a dimensionality of 4 (Zhizhin, 2014 a, b), the unit cell dimension in crystalline beryllium fluoride is also 4.

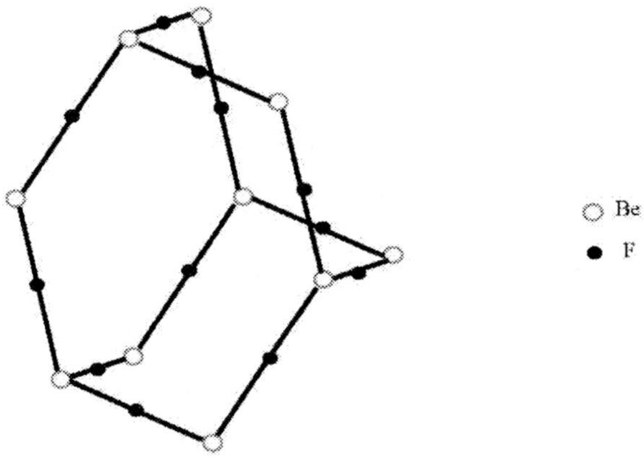

Figure 4.6. The structure of crystalline beryllium fluoride.

If the anomalous elements have one electron in the outer orbital *s* and subshell *d* completely (or almost completely) filled, then the element at the expense of *s* electron forms a linear molecule, such as a linear molecule oxide X_2 O, where X is the anomalous element (Cu, Pd, Ag, Pt, Au, Rg). However, due to the donor-acceptor bond linear molecule can form complex structures in the space. We choose element Cu from second group of anomalous elements. Figure 4.7 shows an exemplary structure formed by linear molecules Cu_2 O.

Each oxygen atom in the structure of Figure 4.7 bonded to four metal atoms (Cu). Two covalent bonds due to the formation of electron pairs divided: one s-electron metal atom and a p-electron atom of oxygen. In addition, there are two more donor-acceptor chemical bond due to the transfer of two electrons from the s-orbital and two electrons from the p-orbitals of the oxygen atom to vacant quantum cell of orbital metal. Thus, oxygen atom acquires valence equal four. In addition, each metal atom is linearly between two oxygen atoms.

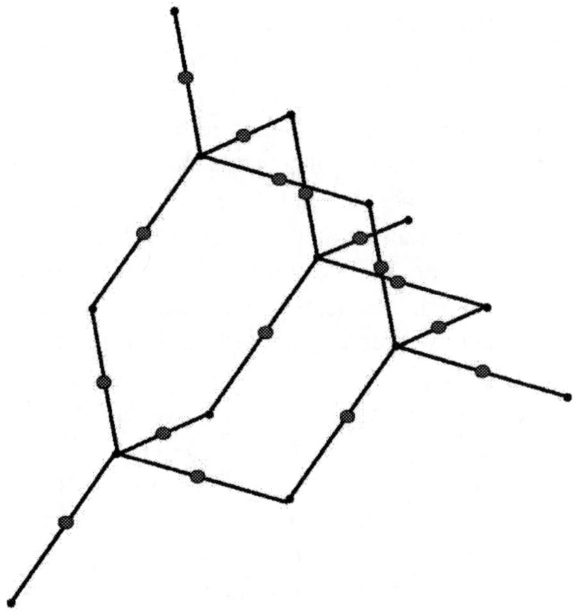

Figure 4.7. The structure of the compound Cu_2O.
A black small circle is oxygen atom. A brown circle is copper atom.

In this structure, oxygen atoms (except oxygen atoms located at the vertices of a cube) form structure is topologically equivalent to the structure of carbon atoms in the molecule of adamantane. As shown in the article of Zhizhin (2014 a) on the basis the monograph Zhizhin (2014 b), the dimension of this molecule is 4. However, the two molecules comprising 10 oxygen atoms have free unallocated space. Therefore, if we set the task of finding the unit cell structure of copper oxide without filling cracks and gaps to help translation the entire space, we need to build polytopic prismahedron (Zhizhin, 2015, 2019), with bases in the form polytopes corresponding to these molecules. Taking a line segment equal to the length of the edge of the cube, in which is inscribed the structure including 10 oxygen atoms, multiply the polytope corresponding to this structure for this segment. We obtain polytopic prismahedron of dimension 5. With this polytopic prismahedron can fill space without gaps and clearances.

Phosphorus, antimony, arsenic and bismuth form already considered polytopes for other elements: a tetrahedron with a center, an octahedron with a center, an adamantane molecule. All these molecules have a dimensionality of 4 or higher. Chlorine, bromine and iodine show the greatest possible numbers of oxidation states, interacting with previously considered elements. The dimensions of these compounds are often higher than three.

A number of elements form chains of tetrahedrons with a center. For example, the element is chrome. It in consequence of the anomalies have one valence electron on the 4 *s*-orbital and five of the electrons on 3*d*-orbital. This allows have of chromium a valence equal 6 in many compounds. Since crystalline chromium oxide CrO_3 consists of chains of tetrahedrons CrO_4, united in two vertices. Each tetrahedron has located in the center of an atom of chromium associated by double bond with each of the four oxygen atoms at the vertices of a tetrahedron. All molecules have the form of a tetrahedron with the center, as shown in (Zhizhin, Diudea, 2016; Zhizhin, Khalaj, Diudea, 2016), have a dimension of 4, i.e., crystalline chromium oxide is a chain of polytopes of dimension 4, united in two vertices. If instead of double bonds are one-time connection with the chromium atom monovalent groups (such as hydroxyl groups), the molecule will be the center of the octahedron, which also would have dimension 4.

A series of binary compounds have a structure in the form of cube with centrum as in titanium chloride at which titanium ions are arranged in the centrum of the cube bat chlorine ions are arranged in vertices of the cube. We will call these structures titanium chloride structure. Haw it is shown in work of Zhizhin, Diudea (2016) this structure has dimension 4.

A series of binary compound have a structure of the mineral rutile TiO_2. In this compound each titanium atom is surrounded by six the oxygen atoms in the octahedral coordination. To compounds with such structure to concern for example fluorides of copper, zinc, magnesium, manganese, cobalt, nickel. We will call these structures rutile structure. In work of Zhizhin, Diudea (2016) it is shown that octahedron with centrum have dimension 4. Therefore, all these structures have dimension 4. A series of binary compounds have structure of wurtzite-mineral ZnS, in which from compound ZnS with the structure of the adamantane zinc atom and sulfur atom have the tetrahedron coordination. The centrum each tetrahedron is vertex of another tetrahedron. The wurtzite structure has compounds ZnO, CdS, ZnS. A series of binary compounds have a fluorite structure – mineral CaF_2 (fluorspar). Each calcium ion in this structure is in cube surrounded by fluorine ions, and each fluorine ion is in tetrahedron surrounded by calcium ions.

4.2.1. The Dimension of the Wurtzite

In the structure of wurtzite every atom of one component has a tetrahedral environment of the atoms of the other component. This results to arrangement of tetrahedrons with center so that vertex one tetrahedron is the center of another tetrahedron we obtain a spatial lattice the unit cell the lattice is a convex shape it is shown in Figure 4.8.

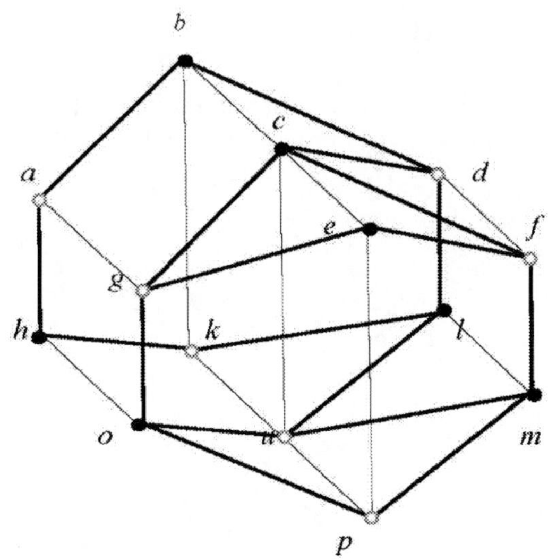

Figure 4.8. The unit cell of the wurtzite.
A white circle is atom of one component. A black circle is atom of another component.

In Figure 4.8, solid stout lines represent chemical bonds of the atoms, and the solid thin lines are only geometric sense outlining contours of the figure. We define the dimension of this figure by the Euler-Poincare equation (4.1). The number of vertices of this figure is equal to 14, i.e., $f_0 = 14$. The number of edges is equal to 29, i.e., $f_1 = 29$. The number of two-dimensional faces is the sum of the number of triangles (8) and number of quadrangles (13), i.e., $f_2 = 21$. The number of three-dimensional faces is equal to 6. It figures are *abcghkon, gceruo, cefump, cdfulm, bcdklu,* and all shape on Figure 4.8 without inner partitions, i.e., $f_3 = 6$. Substituting these values $f_i, (\mathrm{i} = 0,1,2,3)$ in the Euler-Poincare equation (4.1), we find that it is satisfied for $n = 4$

$$14 - 29 + 21 - 6 = 0.$$

Thus, the dimension of polytope on Figure 4.8 is equal to 4, i.e., the unit cell structure of the wurtzite has dimension 4.

4.2.2. The Dimension of the Fluorite

On example of compound $MnCl_2$ we look at the structure of fluorite. Isolate magnesium atoms lying at the centers of the cube faces (●), and chlorine atoms (▲), forming a smaller cube inside the bigger cube, which are located at the vertices of magnesium atoms (Figure 4.9).

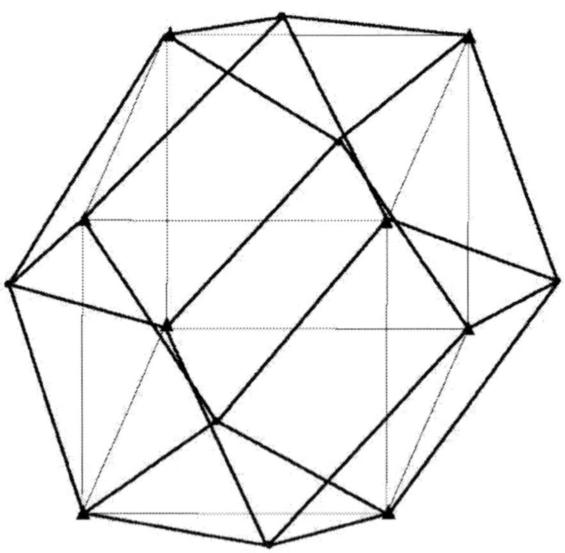

Figure 4.9. The unit cell of the fluorite. ●- magnesium atom, ▲ - chlorine atom.

From Figure 4.9 it follows that the number of vertices is 14, i.e., $f_0 = 14$, the number of edges is 36, i.e., $f_1 = 36$, the number of flat faces is sum from number of triangles (24) and number of rectangles (6) (smaller cube faces), i.e., $f_2 = 30$. The number three-dimension shape to sum up from smaller cube (1), pyramids on its faces (6) and figure (22) without inner parts (1), i.e., $f_3 = 8$ 8. Substituting these values $f_i, (i = 0,1,2,3)$ in the equation (4.1), we find that it is satisfied for $n = 4$

$$14 - 36 + 30 - 8 = 0.$$

Thus, the dimension of polytope on Figure 4.9 is equal to 4, i.e., the unit cell structure of the fluorite has dimension 4. In the conclusion of the segment, we give the table of the most common of binary compounds , indicating the type of structures of the molecules they form.

Table 4.1. Binary compounds of the elements

N	Type of the structure	The compounds of elements with this type of the structure
1	rock salt	LiF, NaF, KF, LiCl, NaCl, KCl, RbCl, LiBr, NaBr, KBr, RbBr, LiI, NaI, KI, MgO, CaO, SrO, BaO, MgS, CaS, SrS, BaS, PbS, MnO, FeO, CoO, NiO, CdO
2	Adamantane	Pb_4O_6, As_4O_6, Sb_4O_6, P_4O_{10}, P_4O_4, $SiCl_4$, BeF_2, HgS, Ag_2O, Cu_2O, ZnS, CuCl
3	titanium chloride	RbF, CsF, CsCl, CsBr, CsI, TiCl
4	Rutile	MgF_2, SnO_2, PbO_2 , CoF_2, NiF_2, CuF_2, ZnF_2, MnO_2, MoO_2
5	Wurtzite	BeO, AlN, ZnO, CdO, ZnS
6	Fluorite	CaF_2, SrF_2, BaF_2, RbF_2, AlF_8, $MgCl_2$, $CaCl_2$, $SrCl_2$, $BaCl_2$, $PbCl_2$, $AlCl_3$, $SiCl_4$, Li_2O, Na_2O, K_2O, Rb_2O, Cs_2O, Li_2S, Na_2S, K_2S, Rb_2S, CdF_2, $MnCl_2$, $FeCl_2$, $CoCl_2$, $NiCl_2$, $ZnCl_2$, $CdCl_2$, Cu_2S

4.3. The Structure of Natural Compounds with a Large Number Types of Atoms

The structure of compounds with a large number of different types of atoms is certainly much more complex than the structure of binary compounds. Consider, for example, gold compounds.

The outer shell of gold atom has one 6*s*-electron and a completely filled 5*d*-orbital. In the compound chlorine triphenylphosphine of gold (Ph_3P)

AuCl gold atom, giving one *s*-electron to chlorine atom, forms ion Au $(Ph_3P)_3^+$ with trigonal coordination (Gillespie, 1972; Perrin, Armarego, Perrin, 1980). Following Zhizhin (2016) we denoted phosphine molecule as a functional group of the compound.

Then ion Au $(Ph_3P)_3^+$ represented in the form of three tetrahedrons with the center, having one common vertex-a gold atom. In the center of each tetrahedron is located phosphorus atom and the remaining vertices of the tetrahedrons are occupied introduced functional groups Ph_3 (Figure 4.10).

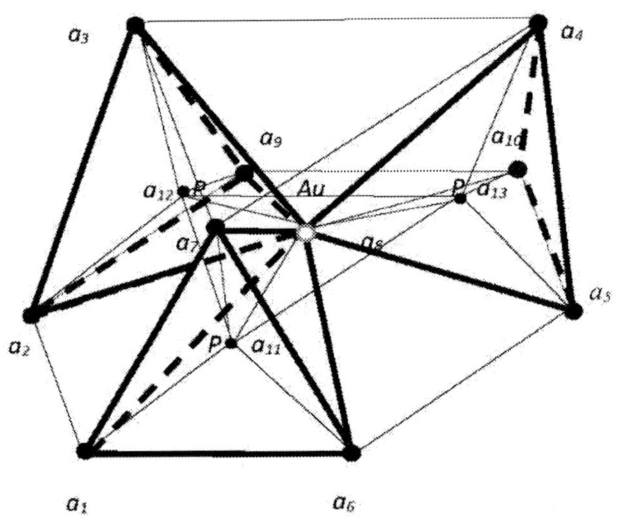

Figure 4.10. Ion Au $(Ph_3P)_3^+$. A white circle is gold atom. A black small circle is phosphorus atom. A black big circle is functional groups Ph_3.

The functional dimension of each tetrahedron with center is still equal to 4. Thus, the ion Au $(Ph_3P)_3^+$ is a collection of three polytopes of dimension 4, having a common vertex. The assertion is proven.

4.3.1. Theorem 4.2

The Ion Au $(Ph_3P)_3+$ has Dimension 5.

4.3.1.1. Proof

To prove the necessity of the three tetrahedrons with a common vertex to form a convex shape. Connect the vertices of a_3, a_4, a_7 by line segments, forming a triangle $a_3\ a_4\ a_7$. Connect also the three centers of the tetrahedrons with each other, forming a triangle $a_7\ a_{11} a_{13}$. Connect the center of the tetrahedrons with vertices corresponding of the tetrahedrons and vertices in the grounds of the tetrahedrons, forming a hexagon $a_1\ a_2\ a_9\ a_{10}\ a_5\ a_6$ (thin lines on Figure 4.5). Define dimension polytope in Figure 4.10 on the Euler-Poincaré equation (Poincaré, 1895) (4.1).

To calculate the number of elements of large dimensions we turn first to a simple polytope, a part of a polytope in Figure 4.10. Temporarily excluded from Figure 4.10 the centers of the tetrahedrons – a_{11}, a_{12}, a_{13}, and all edges emanating from these vertices. Then, the polytope has 13 vertices, i.e., $f_0 = 10$. The number of edges of the polytope is sum the number edges of three tetrahedrons ($6 \bullet 3 = 18$ tetrahedrons at the base figure (3), the number of the edges connecting vertices of the tetrahedrons at top of Figures 4.10. Thus, the number of edges polytope without centers of the tetrahedrons equal 24, i.e. $f_1 = 24$. The number of flat elements is sum of the flat faces of tetrahedrons ($4 \bullet 3 = 12$ hexagon, 3 triangles between tetrahedrons at base of figure, 3 lateral tetragons , 4 triangles of tetrahedron at top of figure. Thus, the number of flat elements is 23, i.e., $f_2 = 23$. The number of three-dimension elements is sum of 4 tetrahedrons, 3 figure between tetrahedrons, 1 hexagon at base and figure composed from boundary flat faces. Thus, the number of three-dimension elements is 9, i.e. $f_3 = 9$. Substituting the values f_i (4.1), we see that it holds for $n = 4$

$$10 - 24 + 23 - 9 = 0.$$

Therefore, three tetrahedrons with common vertices is polytope with dimension 4.

For add centers in tetrahedrons the number of the vertices becomes equal 13, i.e., $(f_0)_c = 13$. For this there add the number of the edges: $4 \bullet 3 = 12$ edges in tetrahedrons with centers, and 3 edges connecting centers. Thus, common number of edges on Figure 4.10 equal 39, i.e., $(f_1)_c = 39$ number flat faces there increases on 18 triangles in the tetrahedrons, 4 triangles in tetrahedron $a_{11}a_{12}a_{13}a_8$, 6 tetragons with vertices part which are centers of the tetrahedrons. Thus, common number of flat faces on Figure 4.10 equal 51, i.e., $(f_2)_c = 51$

dimensions faces increases on $4{\bullet}3 = 12$

centers, tetrahedron $a_{11}a_{12}a_{13}a_8$, figure $a_1\ a_2\ a_9\ a_{10}\ a_5\ a_6\ a_{11}a_{12}a_{13}a_8$, prism $a_{11}a_{12}a_{13}a_3a_4a_7$, 3 pyramids with vertex a_8 ($a_8a_1a_2a_{12}a_{11}$, $a_8a_5a_6a_{11}a_{13}$, $a_8a_9a_{10}a_{12}a_{13}$), 3 prisms ($a_1a_2a_3a_7a_{11}a_{12}$, $a_5a_6a_4a_7a_{11}a_{13}$, $a_9a_{10}a_3a_4a_{12}a_{13}$). Thus, common number of three-dimension faces on Figure 4.10 equal 30, i.e., $(f_3)_c = 30$

It is known from the preceding that the Figure 4.10 has polytopes of dimension 4. Each tetrahedron with center there is polytope of dimension 4 and 3 tetrahedrons without center, but with a common vertex, there is a polytope of dimension 4. In addition, in Figure 4.10 between any two tetrahedrons with the center is polytope dimension 4. Obviously, such polytopes are 3. That to proof this statement we consider any polytope from them. For example, the polytope $a_4a_5a_6a_7a_{11}a_{13}$ (Figure 4.11).

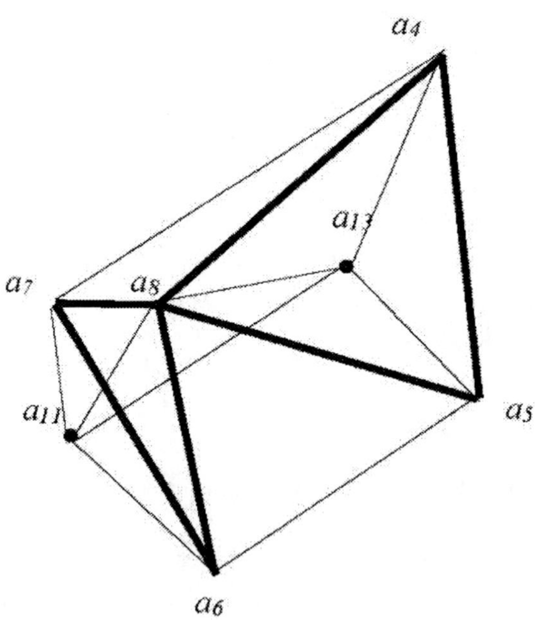

Figure 4.11. The polytope $a_4a_5a_6a_7a_{11}a_{13}$.

Figure 4.11 has 7 vertices $(f_0 = 7)$; 15 edges $a_{11}a_7$, $a_{11}a_8$, $a_{11}a_6$,$a_{11}a_{13}$, a_7a_6, a_7a_8, a_7a_4, a_8a_4, a_8a_{13}, a_8a_5, a_6a_5, $a_{13}a_5$, a_4a_{13}, a_4a_5, a_8a_6; 14 flat faces $a_{11}a_8a_6$, $a_{11}a_7a_8$, $a_6a_7a_8$, $a_{11}a_7a_6$, $a_8a_5a_{13}$, $a_8a_{13}a_4$, $a_5a_{13}a_4$, $a_8a_5a_4$, $a_6a_8a_5$, $a_{11}a_8a_{13}$, $a_7a_8a_4$, $a_6a_{11}a_{13}a_5$, $a_6a_5a_7a_4$, $a_{11}a_7a_{13}a_4$; 6 three-dimension faces $a_6a_{11}a_8a_7$, $a_8a_5a_{13}a_4$, $a_7a_8a_{11}a_{13}a_4$, $a_6a_{11}a_8a_{13}a_5$, $a_{11}a_6a_7a_{13}a_5a_4$, $a_6a_7a_8a_4a_5$. Therefore, for Figure 4.11 are $f_0 = 7, f_1 = 15, f_2 = 14, f_3 = 6$. Substituting the values f_i in equation (4.1), we see that it holds for n = 4

7 – 15 + 14 – 6 = 0.

This proof that Figure 4.11 has dimension 4.

As each figure in polytope on Figure 4.10 is 3, so common number polytopes with dimension 4 in Figure 4.10 equal 7, i.e., $(f_4)_c = 7$ interesting to note that the four-dimensional polytopes included in the polytope on Figure 4.10 have in common both three-dimensional polytopes and a common zero-dimensional vertex face a_8. the values $(f_i)_c$ in equation (4.1), we see that it holds for n = 5

13 – 39 + 51 – 30 + 7 = 2.

Theorem 4.2 is proved.

Even for the chemical bonds of magnesium with valence 2, compounds of higher dimension are formed. Consider a molecule of bis (neopentyl) magnesium $Mg(C_5H_{11})_2$ (Gillespie, Hargittai, 1991). Magnesium in this molecule exhibits a valence of 2. In each group C_5H_{11}, the carbon atoms form the geometric form of a tetrahedron centered. This already gives the dimension of this form equal to 4. In addition, around each carbon atom there is also a tetrahedral coordination of other atoms (hydrogen and carbon). Each group C_5H_{11} can be represented in the form of a tetrahedron with a center in which its vertices contain functional groups CH_3, and in the fourth (attached to the magnesium atom) is a functional group CH_2. At the center of the tetrahedron is a carbon atom. Then the bis (neopentyl) magnesium molecule has the form of two tetrahedrons with a center connected to each other by a magnesium atom (Figure 4.12). Functional groups CH_3 are located in the vertices $a_1, c_1, d_1, a_2, c_2, d_2$ 2 are located in the vertices b_1, b_2; at the points o_1, o_2

magnesium atom.

Valentine bonds are indicated in Figure 4.12 with a brown color. The remaining edges (black) serve to form a convex figure (polytope), the dimension of which must be established.

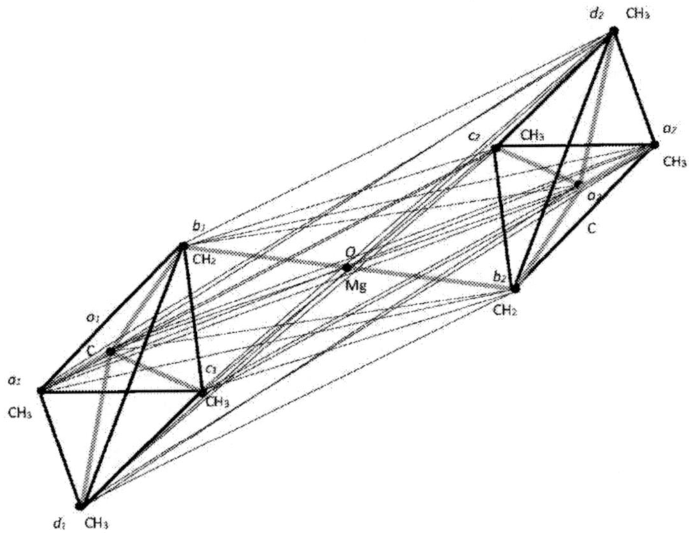

Figure 4.12. The structure of bis(neopentyl) magnesium molecule.

4.3.2. Theorem 4.3

The dimension of bis(neopentyl) magnesium molecule equal to 6.

4.3.2.1. Proof

For proof of theorem 2 we noted that polytope on Figure 4.12 is 5 – cross-polytope with centrum (Figure 4.13).

Comparing Figures 4.13 and 4.12, we see that these figures are topologically equivalent, that is, in Figure 4.13, the same vertices are shown as in Figure 4.12. Moreover, each of the corresponding vertices in Figure 4.13 is incidental to the number of edges as in Figure 4.12 and the connection of vertices by edges in Figure 4.13 is topologically the same as in Figure 4.12. If we denote in Figure 4.12 the edges issuing from vertex O to other vertices in fat black, the remaining figure, as can be seen, is the 5 – cross-polytope, given in the monograph by Zhizhin, (2014 b). In addition, the vertex O is the center of 5 – cross-polytope. As follows from Zhizhin (2014 b) 5 – cross-polytope has 10 vertices ($f_0 = 10$)

(f_1= 40), 80 triangular faces (f_2= 80), 80 tetrahedrons (f_3 –

cross-polytopes (f_4

the 5 – cross-polytope adds, according to Figure 4.13, 10 edges ($oa_1, ob_1, oc_1, od_1, oo_1, ob_2, oa_2, oc_2, od_2, oo_2$

$\{o_1b_1o, b_1a_1o, b_1oa_2, b_1od_2, b_1oc_2, b_1oc_1, c_2d_2o, c_2od_1, c_2oa_2, c_2oo_2, o_2d_2o, o_2oa_2, o_2ob_2, b_2a_2o, b_2od_2, b_2oc_2, b_2oc_1, b_2d_1o, c_1od_1, c_1od_2, c_1oa_2, c_1oo_1, a_1o_1o, o_1d_1o\}$,

28 tetrahedrons ($b_1od_2a_2, b_1c_1oo_1, b_1d_1oa_1, b_1a_1d_2o, b_1od_1a_2, b_1oa_2c_2, b_1od_1c_1, c_2oa_1d_1, c_2b_2oo_2, c_2a_2od_2, c_2a_1d_2o, c_2oa_2d_1, c_2od_1b_1, c_2oa_2b_2, c_1oa_2d_2, c_1a_1od_1, c_1d_1a_2o, c_1oa_1d_2, c_1od_2b_2, c_1od_1b_1, o_1a_1d_1o, b_2od_1a_1, b_2d_2oa_2, b_2a_2d_1o, b_2od_2a_1, b_2oa_1c_1, b_2od_2c_2, o_2d_2a_2o$), 18 4 - simplexes ($b_1a_1od_2c_2, b_1c_1d_1a_1o, b_1o_1a_1oc_2, b_1od_2o_2c_2, b_1od_2a_2o_2, c_1d_1oa_2b_2, c_1o_1d_1b_2o, c_1oa_2o_2b_2, c_1oo_2d_2o_2, c_2b_2a_2d_2o, c_2o_2d_2ob_1, c_2oa_1o_1b_1, c_2oa_1d_1o_1, b_2a_2od_1c_1, b_2od_1o_1c_1, b_2od_1a_1o_1, o_2c_2od_2a_2b_2$), 6 5 - simplexes ($o_1b_1a_1od_1c_1, c_2od_2a_2o_2b_2, c_1od_2a_2o_2b_2, b_2oa_1o_1d_1c_1, c_2od_1a_1o_1b_1, b_1oa_2o_2d_2c_2$.

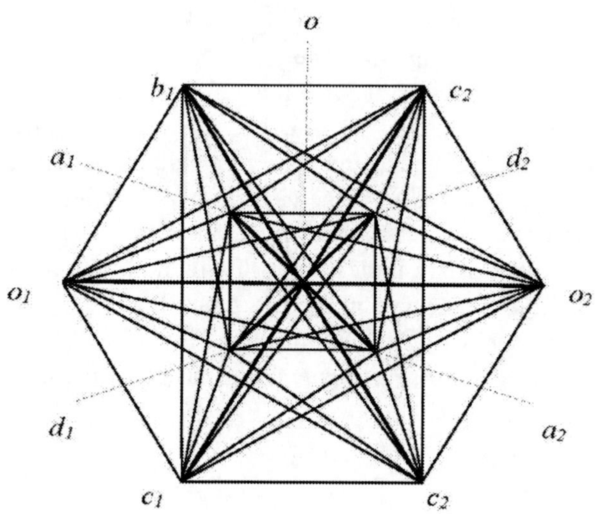

Figure 4.13. The 5 – cross-polytope with centrum.

Adding the obtained quantities of geometric figures of different dimensions connected with the center of the 5 – cross-polytope to the corresponding numbers of figures not connected with the center of the 5 – cross-polytope, we obtain the total number of geometric figures of different dimensions in the 5 – cross-polytope with center: $f_0 = 11, f_1 = 50, f_2 = 104, f_3 = 108,$ $f_4 = 50, f_5 = 7.$ Substituting these values into equation (4.1), we find that the Euler-Poincaré equation is satisfied for $n = 6$

$$11 - 50 + 104 - 108 + 50 - 7 = 0.$$

This proves that a 5 – cross-polytope with center has dimension 6. Theorem 4.3 is proved.

It should be noted that the above evidence accurately lists (in view of the work Zhizhin, 2014 b) all the 108 three-dimensional figures included in the 6-dimensional 5 – cross-polytope with the center.

If the electron pairs of magnesium at the second energy level enter into a chemical bond, then its valence is more than two. For example, in Grignard reagent the magnesium valence is 4 and in the vicinity of magnesium atom there is tetrahedral coordination. While the nearest neighborhood of the magnesium atom has a dimension of 4, and with the account of the attached groups of atoms this dimension is even higher. An interesting example is the complex magnesium ion Mg $(OAsMe_3)_5^{2+}$, Me = CH_3. In this compound, magnesium exhibits a valence of 5. In this case, the nearest environment of magnesium is of dimension 5. Indeed, the nearest environment of magnesium by oxygen atoms has the form of a 4-simplex with a center in the magnesium atom (Figure 4.14). At the vertices *a, b, c, d ,e* of the polytope, in Figure 4.14, there are oxygen atoms, in the vertex *o* there is a magnesium atom. The valence bonds are indicated in Figure 4.14 with a fat black color, the other edges (thin black) are needed to create a convex figure in space. The vertices together with the connecting ribs form a 4-simplex. The addition of a magnesium atom and valence bonds converts this polytope into a 4-simplex with a center.

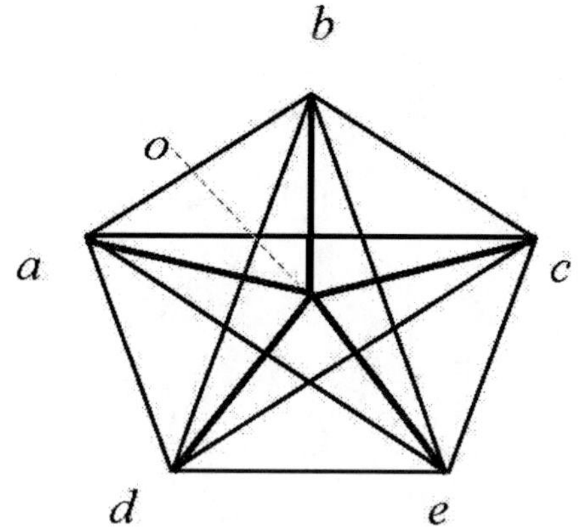

Figure 4.14. The 4 – simplex with centrum.

In Figure 4.14 can indicate 6 vertices (f_0 = 6); 15 edges (f_1

20 trigonal faces (*abc, aeb, abo, abd, bcd, bco, bce, aeo, aed, aec, edo, edc, edb, dco, dca)*, f_2

15 tetrahedrons (*abed, abec, abcd, dbce, aecd, obcd, oecd, aoed, aoeb, aobc, boed, coae, doeb, eobc, aocd*), f_3

6 4 - simplexes (*abcde, abedo, abeco, abcdo,dbceo, aecdo)*, f_4

Substituting values f_i

Poincaré equation is satisfied for $n = 5$

$$6 - 15 + 20 - 15 + 6 = 2.$$

This proves that a 4-simplex with center has dimension 5. If we take into account the presence of other atoms in the ion Mg $(OAsMe_3)_5^{2+}$, then its dimension will be even higher.

Such compounds can form other alkaline-earth elements, i.e., calcium and barium.

Elements Al, Ga, In and Tl have vacant *d*- and *f*-orbitals and tend to supplement their valence shell to 6 electron pairs, and in several compounds In and Tl have more than 6 electron pairs. These elements in many compounds exhibit tetrahedral coordination in the vicinity of the atom. Taking into account the possible addition of other elements to tetrahedral coordination, complex compounds with high dimensionality can arise. For example, aluminum (a biogenic element) forms a cyclic compound $[(CH_3)_2AlF]_4$ (Figure 4.15).

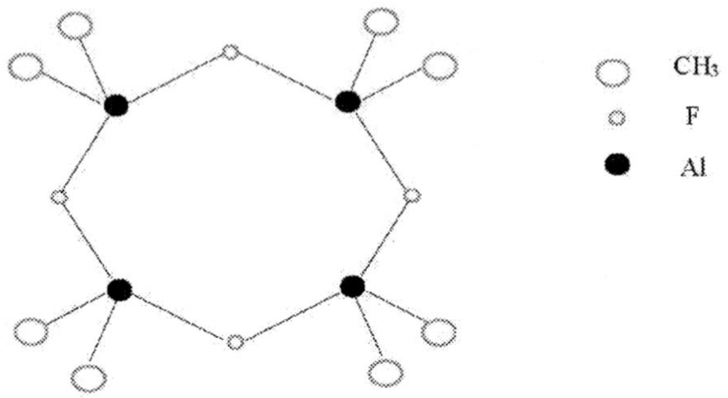

Figure 4.15. A cyclic compound $[(CH_3)_2AlF]_4$.

If we form a convex figure from Figure 4.15, we get the polytope shown in Figure 4.16. At the vertices of a_1, a_4, a_7, a_{10}

the vertices $a_{13}, a_{14}, a_{15}, a_{16}$

CH_3 are located in the $a_2, a_3, a_5, a_6, a_8, a_9, a_{11}, a_{12}$

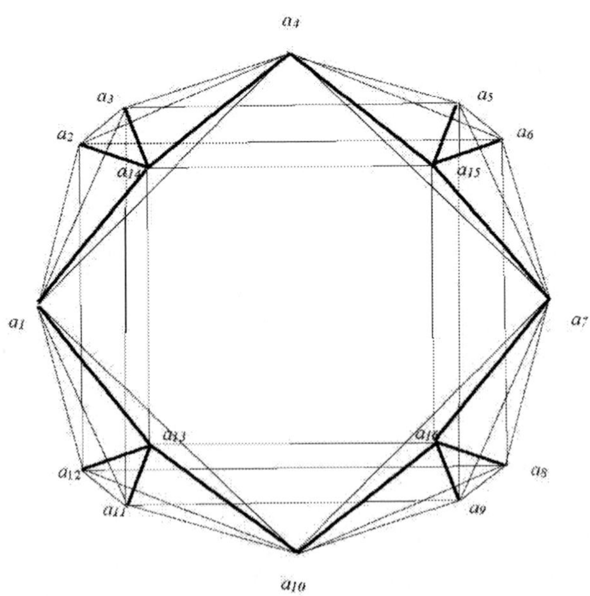

Figure 4.16. The convex polytope of cyclic compound $[(CH_3)_2AlF]_4$.

The thick black lines in Figure 4.16 represent valence bonds, the thin black lines correspond to ribs that have spatial significance.

4.3.3. Theorem 4.4

The polytope of cyclic compound $[(CH_3)_2AlF]_4$ has dimension 5.

4.3.3.1. Proof

The polytope in Figure 4.16 has 16 vertices, (f_0 = 16); 52 edges, (f_1

In addition, it has 4 polytopes of dimension 4 each (tetrahedrons with a center) $a_1a_2a_3a_4a_{14}, a_4a_5a_6a_7a_{15}$, $a_7a_8a_9a_{10}a_{16}, a_{10}a_{11}a_{12}a_{13}$ with a center has 10 triangular faces. This gives 40 triangular faces in the polytope 10. In addition, three triangular faces are formed at the vertices a_1, a_4, a_7, a_{10} with horizontal and vertical sides. This gives another 4 · 3 = 12 triangles. There are 4 more rectangular faces ($a_{13}a_{14}a_{15}a_{16}, a_6a_2a_8a_{12}, a_{11}a_5a_3a_9, a_1a_7a_{10}a_4$

($a_3a_5a_{15}a_{14}, a_3a_2a_5a_6, a_{15}a_2a_6a_{14}, a_6a_8a_{15}a_{16}, a_6a_8a_5a_9, a_{15}a_5a_9a_{16}$,

$a_{11}a_9a_8a_{12}, a_9a_{11}a_{13}a_{16}$, $a_{12}a_8a_{13}a_{16}$, $a_2a_{12}a_{14}a_{13}, a_{12}a_2a_3a_{11}, a_3a_{11}a_{13}a_{14}$).

Thus, the total number of two-dimensional faces 68, f_2 = 68. Each tetrahedron with a center has 5 tetrahedrons. Therefore, the total number of tetrahedrons in Figure 4.16 is 5 · 4 = 20. Each of the vertices a_3, a_7, a_4, a_{10} the vertex of the three pyramids. The total number of these pyramids is 12: $a_5a_{15}a_3a_4a_{14}, a_6a_2a_3a_4a_5, a_4a_2a_6a_{15}a_{14}, a_1a_2a_3a_{12}a_{11}$, $a_1a_{14}a_3a_{11}a_{13}$, $a_{10}a_{12}a_{11}a_8a_9, a_{10}a_{12}a_{13}a_8a_{16}, a_{10}a_{11}a_{13}a_{16}a_9, a_7a_8a_9a_5a_6, a_7a_9a_{16}a_5a_{15}$, $a_7a_8a_{16}a_6a_{15}$. There are four triangular prisms: $a_2a_3a_{14}a_5a_6a_{15}, a_{15}a_5a_6a_8a_9a_{16}, a_{11}a_{12}a_{13}a_8a_9a_{16}, a_2a_3a_{14}a_{11}a_{12}a_{13}$, and six quadrangular prisms: $a_{13}a_{14}a_{15}a_{16}a_3a_5a_9a_{11}, a_{13}a_{14}a_{15}a_{16}a_2a_6a_8a_{12}, a_{13}a_{14}a_{15}a_{16}a_1a_4a_7a_{10}$, $a_2a_6a_8a_{12}a_3a_5a_9a_{11}, a_1a_2a_4a_6a_7a_8a_{10}a_{12}, a_1a_3a_4a_5a_7a_9a_{10}a_{11}$. Then the total number of three -dimensional figures is 42, f_3 = 42.

In addition to the 4 tetrahedrons mentioned with the center, as four-dimensional figures, there are another four-dimensional figures. In particular, this is a figure (*F*), shown in Figure 4.17. Indeed, this figure has 12 vertices, f_0 (F) = 12; 24 edges, $f_1(F)$ = 24; 19 two-dimensional faces, f_2 F) = 19; and 7 three-dimensional figures, f_3 F) = 7. Substituting these values into the Euler-Poincaré equation (4.1), we obtain that it is satisfied for $n = 4$

$$12 - 24 + 19 - 7 = 0.$$

This proofs that polytope F has dimension 4.

Four identical polytopes of dimension 4 exist in a neighborhood of each of the vertices a_1, a_4, a_7, a_{10}. L) is depicted in Figure 4.18. It has 7 vertices, f_0 (L) = 7; 15 edges, f_1 L) = 15; 14 two-dimensional faces, f_2 (L) =14; and 6 3D facets, f_3 L) = 6. Substituting these values into the Euler-Poincaré equation (4.1), we obtain

$$7 - 15 + 4 - 6 = 0,$$

i.e., the equation (4.1) holds for $n = 4$ and all the polytopes L has dimension 4.

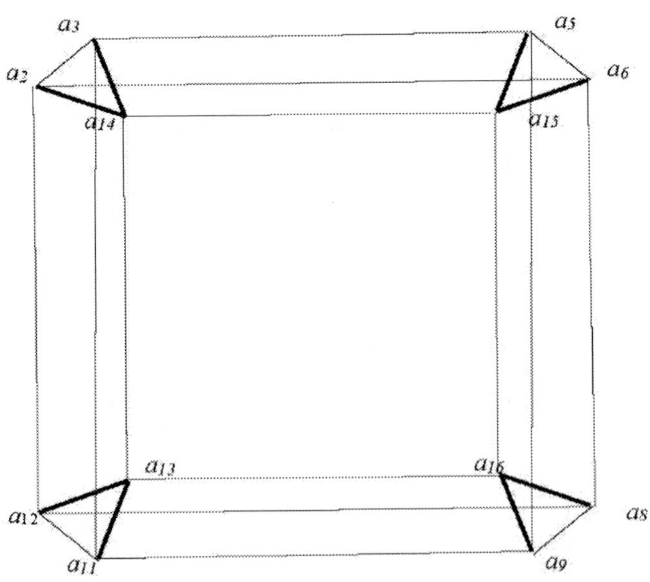

Figure 4.17. The 4 – dimension polytope F included in Figure 4.16.

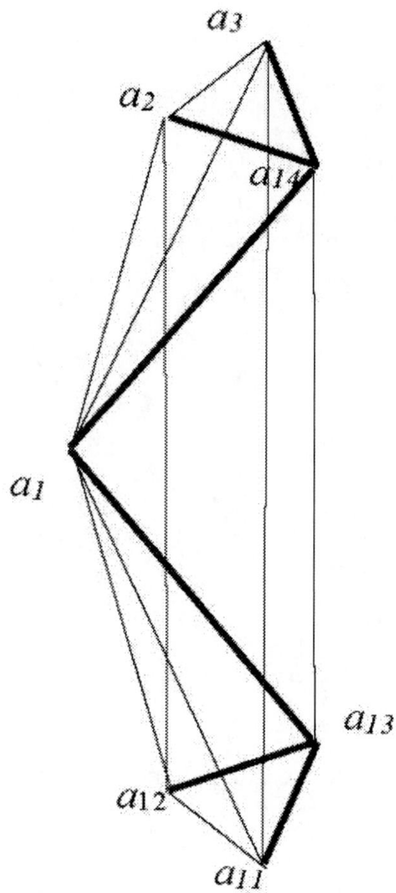

Figure 4.18. The 4 – dimension polytope L included in Figure 4.16.

Three more topologically equivalent polytopes of dimension 4 can be distinguished from Figure 4.16. Each of these polytopes consists of a rectangular prism and four tetrahedrons connected to each other in a cycle along the vertices of a_1, a_4, a_7, a_{10}

$$a_1a_4a_3a_{14}a_5a_{15}a_7a_{16}a_8a_{10}a_{13}a_{12}, a_1a_2a_{14}a_4a_{15}a_6a_7a_{16}a_9a_{10}a_{11}a_{13}, a_1a_2a_3a_4a_5a_6a_7a_8a_9a_{10}a_{11}a_{12}$$

One of them (polytope *K*) is shown in Figure 4.19.

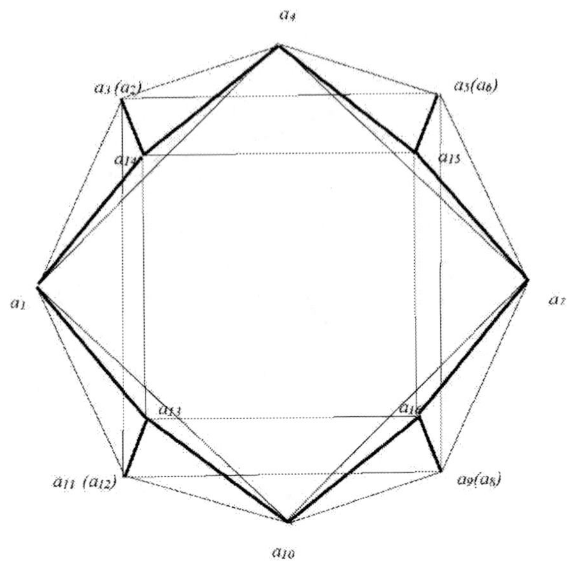

Figure 4.19. The 4 – dimension polytope K included in Figure 4.16.

The K polytope has 12 vertices, $f_0(K) = 12$; 32 edges, $f_1(K) = 32$; 31 two-dimensional faces, f_2 (K) = 31; and 11 3D facets, f_3 $K)$ = 11. Substituting these values into the Euler-Poincaré equation (4.1), we obtain

$$12 - 32 + 31 - 11 = 0,$$

i.e., the equation (4.1) holds for n = 4 and all the polytopes K has dimension 4.

Thus, the polytope on Figure 4.16 has 11 polytopes of dimension 4. Therefore, for polytope on Figure 4.16 the Euler-Poincaré equation (4.1) has face

$$16 - 52 + 68 - 42 + 11 = 2,$$

i.e., it hold for $n = 5$. This proofs theorem 4.4.

4.4. Pomegranate Texture

Garnet is one of the oldest precious stones. They are widely used in many fields of science and technology. The general formula of pomegranates is depicted as $A_3B_2(SiO_4)_3$ A, you can substitute the following elements: calcium, magnesium. iron, manganese. Instead of the letter B, you can substitute aluminum, iron, chrome. Pomegranates of different compositions are called differently. For example, a garnet described by the formula $Mg_3Al_2(SiO_4)_3$

are observed in pyrope: a tetrahedron with a center, an octahedron with a center, and a Thomson cube with a center (Belov, 1976; Akhmetov, 1990). According to academician N.V. Belov, these components are the main structural motives of any garnet. A tetrahedron with a center and an octahedron with a center have already been discussed in the previous chapters. Their dimension is 4. Let us determine the dimension of the Thomson cube with the center (Figure 4.20).

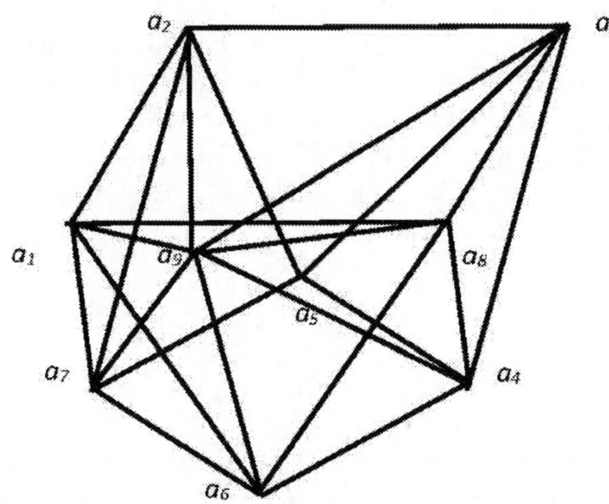

Figure 4.20. The Thomson cube with the center.

Figure 4.20 has 9 vertices $(f_0 = 9)$; 24 edges $(f_1 = 24)$ $a_1a_2, a_1a_8, a_1a_9, a_1a_6, a_1a_7, a_2a_3, a_2a_5, a_2a_9, a_2a_7, a_3a_4, a_3a_8, a_3a_9, a_3a_5, a_4a_8, a_4a_5, a_4a_9,$; 25 flat faces $a_4a_6, a_5a_9, a_5a_7, a_6a_7, a_6a_8, a_6a_9, a_7a_9, a_8a_9$

$(f_2 = 25)$ $a_1a_2a_3a_8, a_1a_2a_9, a_9a_1a_7, a_2a_1a_7, a_1a_8a_6, a_1a_9a_6, a_1a_9a_7, a_1a_7a_6, a_2a_9a_7, a_2a_9a_5, a_2a_3a_9, a_2a_3a_5, a_3a_5a_9,$; 6 three- $a_3a_8a_9, a_3a_8a_4, a_3a_4a_9, a_4a_8a_9, a_4a_6a_9, a_4a_6a_8, a_4a_5a_9, a_4a_5a_6a_7, a_5a_7a_9, a_6a_8a_9, a_6a_7a_9, a_1a_8a_9$

dimension faces $(f_3 = 10)$ $a_1a_2a_3a_8a_9, a_1a_2a_7a_9, a_1a_7a_9a_6, a_1a_8a_6a_9, a_2a_7a_7a_9, a_3a_2a_5a_9, a_3a_9a_4a_8, a_4a_8a_6a_9,$. $a_4a_6a_5a_9a_7, a_1a_2a_3a_8a_4a_5a_7a_6$

Substituting the values f_i $n =$

4

$$9 - 24 + 25 - 10 = 0.$$

This proof that Figure 4.20 has dimension 4.

This proves that the dimension of the Thomson cube with the center is 4. Thus, the structure of the garnet must consist of adjacent to each other the listed four-dimensional geometric shapes. Four-dimensional geometric shapes, when filling the space, must contact over three -dimensional shapes. This does not exclude the possibility of four-dimensional shapes in the structure having common elements and other dimensions (two-dimensional face, edge or vertex). But without the contact of four-dimensional figures, a continuous filling of space cannot be formed from three-dimensional figures. Of the listed figures with a center (Thomson's cube, octahedron, tetrahedron), it remains unclear how the structure of the garnet is formed, which three-dimensional figures are common between these figures. In the well -known works on the structure of garnet, this problem was not solved.

Thus, despite the long acquaintance with the garnet gemstone, its structure remains unknown. This is all the true for a mixture of gemstones such as garnets.

Conclusion

The study showed that inanimate (inert) substances in nature have a number of valuable properties. Self-organization processes take place in them, allowing the creation of perfect structures, including the structures of precious stones. These structures are so complex, albeit perfect, that a person cannot describe these structures, being within the framework of standard established ideas. Mathematical analysis has shown that all these structures have the highest dimension. Moreover, the greater the number of different types of atoms involved in the formation of the structure, the higher the dimension of the structure. The structures of inert substances are characterized by scaling processes (Kadanoff, 1966), i.e., spontaneous increase in the scale of elementary geometric units. An important manifestation of the activity of the so-called inert matter is the formation of quasicrystals in inter-metallics, the translational order in which is found only in the space of the highest dimension (Zhizhin, 2014 c; Zhizhin, Diudea, 2016; Shevchenko, Zhizhin, Mackay, 2013; Zhizhin, 2018). The most striking example of self-organization is the garnet gem. This stone has been known to man for over 1000 years, but as shown by mathematical analysis, its structure remains unknown. At present, it cannot be described even using the concept of the highest dimension of the structure space. All this testifies that the substance of inanimate nature creates dynamic self-governing systems (Zhizhin, 2005) and the term "inert," introduced by V.I. Vernadsky (Vernadsky, 1912, 1977) for their designation does not correspond to their properties. They have many characteristics of living matter, although they do not contain biomolecules.

Chapter 5

“Inert” Substances as a Self-Regulating Medium Tending to Capture Space

Abstract

Existing methods for the distribution of molecules of “inert” matter in space due to the expansion of the geometric shape of the molecules and the chemical bond between the initial molecule and larger shells are considered. Such methods are based on attaching the geometric shapes of the original molecule to each other along whole faces of different dimension or attaching polytopic prisms to the faces of the original molecule. Analytical expressions are obtained that determine the resulting geometric figures, which are clusters of increasing size. It is shown that polytopes of higher dimension obtained as a result of expansion of molecules of “inert” substance form new classes of polytopes of higher dimension that do not belong to polytopes of the simplex, cube, and cross-polytope types.

Introduction

In recent decades, interest in the study of dissipative systems has sharply increased (Zhizhin, 2005), i.e., systems that support their existence due to self-organization and consumption (dissipation) of energy entering the system from the environment, external in relation to the system. As a result of the formation of a dissipative system, an inhomogeneous distribution of matter over space is created-a spatial dissipative structure is formed. Attempts to describe dissipative systems scientifically on the basis of systems of differential equations are associated with serious mathematical difficulties. Here, as a first step, you can limit yourself to the geometric side of the issue. In this regard, the problem arises of filling the space with geometric figures (Zhizhin, 2010). Moreover, to reflect the process of increasing the dissipative structure, it was proposed to introduce the concept of a growing geometric variety. It is essential that dissipative systems can arise in both living and inanimate nature. According to V.I. Vernadsky

(Vernadsky, 1925, 1977), a characteristic feature of living matter is its desire, due to reproduction, to seize the territory adjacent to its location. However, the so-called "inert" substance also has this property. In this case, as well as for living matter, the implementation of this process requires the fulfillment of at least two conditions: the presence of "food" products and the corresponding circumstances suitable for the reproduction or attachment of molecules of the substance. In this chapter, we will consider geometric models of growing dissipative systems of inanimate nature.

5.1. Geometric Growth Models of Dissipative Systems

As a variant of representing a growing geometric manifold, the following method of plane tiling was proposed. Having some regular polygon on the plane, we extend the sides of the polygon to the area outside the polygon until the nearest extensions intersect with each other. We connect the formed vertices with each other by segments, like the sides of a new polygon, similar to the initial one. Then we will do the same operation with a new polygon, etc. In this case, an unlimited increase in the polygon and the triangles included in it, formed from the intersection of the sides, occurs. The area inside the start polygon is also filled with similar polygons of ever decreasing size. Discreet expansion of the polygons with respect to their common center occurs. All the same, the figures have the same dimension (2), but different sizes. Figure 5.1 shows an example of constructing such a geometric manifold in the case of a regular pentagon.

The disadvantage of this model is a sharp increase in the distance between the vertices of the polygon with distance from the center of the polygons. If we assume that atoms or molecules of a chemical compound are located at the vertices, then such an increase in distance would inevitably lead to the rupture of chemical bonds.

An interesting continuation of the construction of geometric manifolds can be considered the diffractograms of intermetallic compounds (Zhizhin, 2014 a, b, c, d). From the analysis of these diffractograms (Zang, Kelton, 1993; Nyman, Andersson, 1979; Mukhopadhyay, et al., 1993) for various intermetallic compounds (quasicrystals), it follows (Zhizhin, 2014 a; Shevchenko, Zhizhin, Mackay, 2013 a, b) that in the vicinity of each luminous spot of the diffractogram as a center, the vertices of expanding pentagons can be seen. Since each point of the diffractogram is of different intensities of the glowing spot of the diffractogram, this means that the expansion of the pentagon occurs from each point of the diffractogram. In essence, we see a picture of a uniform expansion of space, displayed by the diffractogram. In this case, there is no break between the vertices of the pentagons and there is no break of chemical bonds. The construction of such geometric manifolds made it possible to detect translational symmetry in quasicrystals in a space of higher dimension, since the existence of an elementary cell of higher dimension was established in these manifolds.

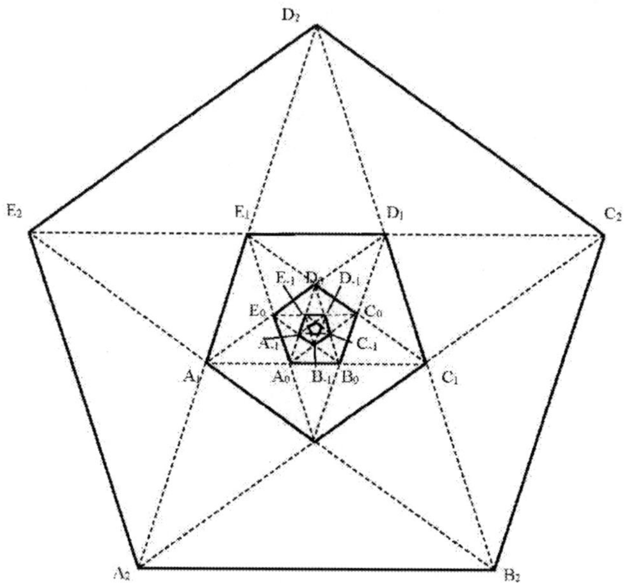

Figure 5.1. Geometric manifold based on a regular pentagon.

There is another way to avoid increasing the distance between vertices indefinitely when constructing a geometric manifold. It follows from the previous chapters that many chemical compounds of "inert" substances are characterized by tetrahedral coordination. Let's say a molecule of some substance has the shape of a tetrahedron. If it is possible to connect other molecules or atoms to the vertices of the tetrahedron, then at least one more edge will come from each of the vertices of the tetrahedron. An example of such a compound would be a boron chloride molecule (Zhizhin, 2018). Connecting the free vertices of these edges, the length of which corresponds to the length of chemical bonds, with new edges, we get a second larger tetrahedron, which contains the original tetrahedron inside it (Figure 5.2).

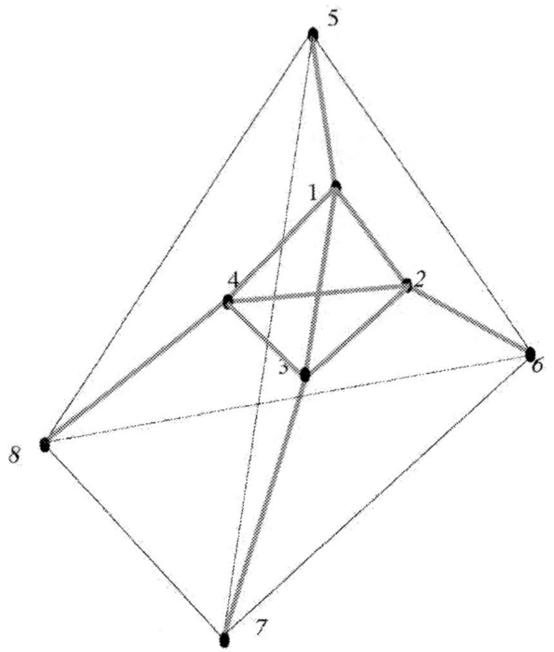

Figure 5.2. The structure of the molecules B_4Cl_4.

5.1.1. Theorem 5.1

The B_4Cl_4 Molecule has Dimension 4.

5.1.1.1. Proof

In Figure 5.2 there is eight vertices, $f_0 = 8$. The number of edges is 16 (8 - 5, 5 - 6, 6 – 7, 8 – 7, 5 – 7, 8 – 6, 4 – 1, 1 – 2, 3 – 2, 1 – 3, 4 – 2, 3 – 4, 5 – 1, 2 – 6, 3 – 7, 8 – 4,), f_1 16. The number of elements of dimension 2 is 14 (triangles 8 – 5 – 7 , 5 – 6 – 7, 8 – 5 – 6 , 8 – 6 – 7, 4 – 1 – 3, *1 – 3 – 2,* 4 – 1 – 2, 4 – 2 – 3 and quadrangles 8 – 4 – 1 – 5, 8 – 4 – 3 – 7, 1 – 5 – 3 – 7, 1 – 5 – 2 – 6, 3 – 2 – 6 – 7, 1 – 2 – 5 – 6), $f_2 = 14$. The number of elements of dimension 3 is 6 (tetrahedrons 8 – 5 – 6 – 7, 4 – 1 – 3 – 2 and prismatoides 8 – 5 – 1 – 4 – 3 – 7, 1 – 5 – 3 – 7 – 2 – 6, 8 – 4 – 1 – 5 – 2 – 6, 4 – 3 – 2 – 8 – 7 – 6), $f_3 = 6$. On Figure 5.2 the edges correspondent of chemical bounds is indicated brown, remain edges (black) it is need for creating convex body.

Substituting the values of the number of elements of different dimension in the equation Euler – Poincaré (4.1), we obtain

$$8 - 16 + 14 - 6 = 0.$$

We find that it holds for $n = 4$. This proves that the figure who projection is shown in Figure 5.2 there is polytope of dimension 4. This proves theorem 5.1.

Note that this is a higher-dimensional polytope of a new type; it does not belong to the known higher-dimensional polytopes of the simplex, cube, cross-polytope types.

Let us assume that to the vertices of the larger tetrahedron, formed at the first step of filling the space with the tetrahedron, there is the possibility of connecting the following atoms or molecules of the "inert" substance. Geometrically, this will correspond to the exit from the vertices of the larger tetrahedron into the surrounding space along one more edge (Figure 5.3).

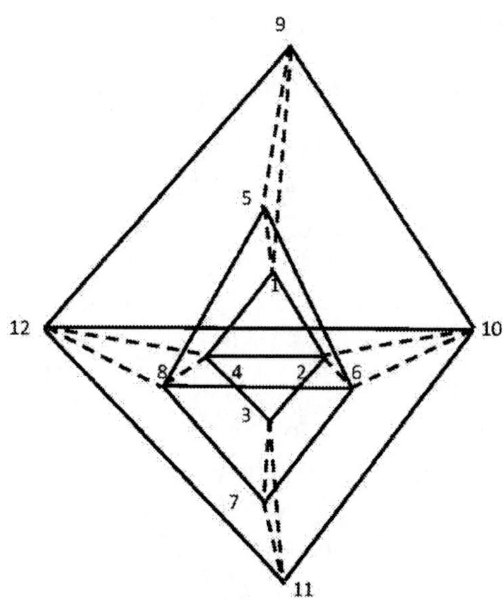

Figure 5.3. The structure of the compound three tetrahedrons with a common center.

Connecting again the free vertices of these edges with edges, we get an even larger tetrahedron. To be able to calculate the dimension of the resulting figure from three tetrahedrons, it is necessary that the same number of edges emanate from each vertex of this figure (Zhizhin, 2019 a). At this stage of construction, each vertex of the largest tetrahedron and each vertex of the smallest tetrahedron are incidentally 4 edges, and each vertex of the middle tetrahedron is incidentally 5 edges. Thus, to introduce the homogeneity of all vertices of the resulting construction, it is necessary to connect an edge each vertex of the smallest tetrahedron with the corresponding vertex of the largest tetrahedron (Figure 5.3). Let's calculate the hanging of elements of different dimensions in the resulting polytope. The number of vertices in Figure 5.3 is 12 ($f_0 = 12$). The number of edges in Figure 5.3 is 30 ($f_1 = 30$): 1 – 2, 1 – 5, 1 – 9, 1 – 3, 1 – 4, 2 – 6, 2 – 3, 2 – 4, 2 – 10, 3 – 4, 3 – 7, 3 – 11, 4 – 8, 4 – 12, 5 – 6, 5 – 7, 5 – 8, 5 – 9, 6 – 7, 6 – 8, 6 – 10, 7 – 8, 7 – 11, 8 – 12, 9 – 10, 9 – 11, 9 – 12, 10 – 11, 10 – 12, 11 – 12.

Let's count the number of elements of dimension 2 in Figure 5.3: 1 – 2 – 3, 1 – 2 – 4, 1 – 2 – 5 – 6, 1 – 5 – 4 – 8, 1 – 3 – 5 – 7, 1 – 5 – 9, 1 – 3 – 9 – 11, 1 – 2 – 9 – 10, 1 – 4 – 9 – 12, 2 – 3 – 4, 2 – 3 – 6 – 7, 2 – 3 – 10 – 11, 2 – 6 – 10, 2 – 4 – 6 – 8, 2 – 4 – 10 – 12, 3 – 4 – 7 – 8, 3 – 4 – 11 -12, 3 – 7 – 11, 4 – 8 – 12, 5 – 6 – 7, 5 – 7 – 8, 5 – 6 – 8, 5 – 6 – 9 – 10, 5 – 8 – 9 – 12, 5 – 7 – 9 -11, 6 – 7 – 10 – 11, 6 – 7 – 8, 7 – 8 – 10 – 11, 9 – 10 – 12, 9 – 10 – 11, 9 – 11 – 12, 10 – 11 – 12, 8 – 6 – 10 – 12, ($f_2 = 34$).

Let's count the number of elements of dimension 3 in Figure 5.3: 1 – 2 – 3 – 4, 5 – 6 – 7 – 8, 9 – 10 – 11 – 12, 1 – 2 – 3 – 5 – 6 – 7, 1 – 2 – 3 – 9 – 10 – 11, 1 – 2 – 5 – 6 – 10, 1 – 3 – 4 – 5 – 7 – 8, 1 – 3 – 4 – 9 – 11 – 12, 1 – 3 – 5 – 7 – 9 – 11, 1 – 2 – 4 – 5 – 6 – 8, 1 – 2 – 4 – 9 – 10 – 12, 1 – 4 – 5 – 8 – 9 – 12, 2 – 3 – 4 – 6 – 8 – 7 , 2 – 3 – 4 – 10 – 11 – 12, 4 – 3 – 7 – 8 – 11 – 12 , 3 – 2 – 6 – 7 – 10 – 11, 4 – 2 – 8 – 6 – 10 – 12, ($f_3 = 17$).

Figure 5.3 has three polytopes of dimension 4: a tetrahedron 1 – 2 – 3 – 4 in a tetrahedron 5 – 6 – 7 – 8, a tetrahedron 5 – 6 – 7 – 8 in a tetrahedron 9 – 10 – 11 - 12, a tetrahedron 1 – 2 – 3 – 4 in a tetrahedron, 9 – 10 – 11 – 12, i.e., $f_4 = 3$. Substituting the obtained values $f_i (i = 0, 1, 2, 3, 4)$ into the Euler-Poincaré equation (4.1), we obtain

$$12 - 30 + 34 - 17 + 3 = 2.$$

Therefore, the dimension of the figure in Figure 5.3 is 5.

Further steps of filling (conquering) space with a tetrahedron will also be accompanied by an increase in the dimension of the figure by one line at each step.

Due to the interaction of electron pairs of several atoms, formation of other compounds is also possible. For example, in Figure 5.4. The image of the B_6Cl_6 molecule is shown. Here, also, the edges corresponding to the chemical bonds is indicated in brown, the remaining edges are necessary for obtaining a convex figure.

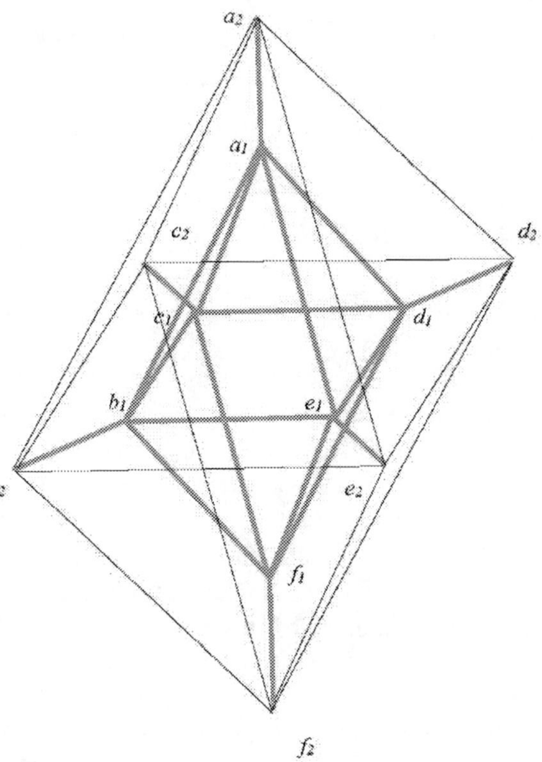

Figure 5.4. The structure of the B_6Cl_6 molecule.

In the compound, both the boron atoms and the chlorine atoms have octahedral coordination. The effective valence of boron in this compound is 5. In the polytope in Figure 5.4, boron atoms are located at the vertices $a_1, b_1, c_1, d_1, e_1, f_1$ and chlorine atoms are located at the vertices $a_2, b_2, c_2, d_2, e_2, f_2$.

5.1.2. Theorem 5.2

The B_6Cl_6 Molecule has Dimension 4.

5.1.2.1. Proof

In this case the number of elements of zero dimension is f_0 = 12. The number of elements of dimension one is f_1 = 12+ 12+ 6 = 30. The number of elements of dimension 2 is sum of the number small triangles 8 and big triangles 8, add 12 quadrangles, i.e., f_2 = 28. The number of elements of dimension 3 is sum two octahedrons and 8 prisms, i.e., f_3
the values of numbers of elements of different dimensions in the equation (4.1), we obtain

$$12 - 30 + 28 - 10 = 0.$$

We find that it holds for $n = 4$. This proves that the figure 5.4 is polytope of dimension 4. This proves theorem 5.2.

The tetrahedron and octahedron are only two representatives of Plato's solids. With each of the regular polyhedron included in the set of Plato's solids, procedures for a discrete increase in their size as they move away from their common center can be considered. The physical implementation of such procedures are the processes of formation of clusters of chemical compounds. We will now formulate and prove a general theorem on the dimension of such clusters for all Platonic solids, if there is or is not a material particle (atom or molecule) in the center of the cluster. Separate problems about clusters for different chemical compounds were considered earlier (Bergman, et al., 1952, 1957; Eiji, A. et al., 2004; Gubin, 1987; Mackay, 1962; Mukhopadhyay, et al., 1993; Samson, 1962, 1964, 1965, 1967, 1972).

5.2. The Dimension of Clusters of Several Shells in the Form of Plato's Bodies

Suppose there is a cluster, each shell of which is a convex regular three-dimensional polyhedron (Plato's body) with some possible number of vertices n and the same for all shells of this cluster. The flat sides of the shells are regular *m*-corner. All shells of this cluster differ only in size and have a common center. Assume that the number of flat edges in the shell is *j*. One denoted S_j a shell satisfying the indicated conditions. Then the following statement is true (Zhizhin, 2019 b, c; 2021).

5.2.1. Theorem 5.3

The dimension *d* of a cluster of *N* shells with a common center is *N* + 2, if there is no atom in the common center, and is equal to *N* + 3 if there is an atom in the common center.

5.2.1.1. Proof

Denote by the symbol *t* the number of edges emanating from each vertex of the shell S_j. Then the total number of edges of the shell is $nt/2$, and the number of faces of the shell is $j = nt/m$ *j* in Plato's body also gives the form of the shell, i.e., sets the numbers *n, m, t* (Table 5.1).

Table 5.1. Defining Plato's bodies by the number of two-dimensional faces of the outer shell

J	*n*	*m*	*t*	*Polyhedron*
4	4	3	3	Tetrahedron
6	8	4	3	Cube
8	6	3	4	Octahedron
12	20	5	3	Dodecahedron
20	12	3	5	Icosahedron

Let a shell with the number of faces *j* and be given. Therefore, the number of vertices n_j, the number of sides m_j at the flat face, the number of edges t_j emanating from each vertex are given. Consider two shells of different sizes with a common center, arranged so that every two corresponding vertices of both shells are on the same straight line connecting them with a common center (there is no atom in the center). Then the space between the shells is filled with prisms, the bases of which are flat faces of the larger and smaller shells. The number of these prisms is equal to the number of faces of the shell, i.e., equal *j*. The number of vertices in a polytope of two shells is $2n_j = f_0$

taking into account the edges connecting the corresponding vertices of the shells, is $n_j t_j + n_j = f_1$

twice the number of flat edges in each shell and the number of edges in one of the shells, i.e., $2j + \frac{n_j t_j}{2} = f_2$

the polytope is equal to $j + 2 = f_3$. Substitute the numbers f_0, f_1, f_2, f_3 the equation of Euler – Poincaré (Poincaré, 1895)

$$\sum_{i=0}^{d-1} (-1)^i f_i(P) = 1 + (-1)^{d-1} \tag{5.1}$$

In (5.1) $f_i(P)$ i in polytope P with dimension d.

Then we have

$$2n_j - n_j(t_j + 1) + 2j + \frac{n_j t_j}{2} - j - 2 = (n_j - \frac{n_j t_j}{2} + j) - 2 = 0.$$

There $(n_j - \frac{n_j t_j}{2} + j)$

dimension 3.

This proves that a polytope composed of two shells S_j with a common center (in the absence of an atom in the center) has dimension 4 for any possible j.

If the cluster consists of three shells S_j with a common center, then $f_0 = 3n_j$. The number of edges in this cluster, taking into account the edges connecting of the corresponding vertices in shells S_j, is equal to $f_1 = 3\frac{n_j t_j}{2} + 2n_j$. The number of two-dimensional faces is $3j + n_j t_j = f_2$

number of three-dimensional faces is $2j + 3 = f_3$

dimensional faces is $f_4 = C_3^2 = 3$. Substitute these numbers in the equation (5.1) Euler-Poincaré

$$3n_j - n_j(\frac{3}{2}t_j + 2) + 3j + n_j t_j - (2\,\mathrm{j} + 3) + 3 = n_j - \frac{n_j t_j}{2} + j = 2.$$

This proves that dimension of the cluster of tree shells S_j is equal 5.

One can write a general expression for the numbers of elements of different dimensions in a cluster with an arbitrary number N of shells S_j and get a formula for calculating its dimension. So, in the N-shell cluster of shells S_j, we have

$$f_0 = Nn_j, f_1 = N\frac{n_j t_j}{2} + (N-1)n_j, f_2 = Nj + (N-1)\frac{n_j t_j}{2}, f_3 = (N-1)\,\mathrm{j} + N,$$
$$f_4 = C_N^2, f_5 = C_N^3, \ldots, f_{N+1} = C_N^{N-1}. \tag{5.2}$$

Substituting the values (5.2) in the equation (5.1) and opening the brackets, one can see that in this case the left side of the Euler-Poincaré equation (5.1) takes the form

$$\sum_{i=0}^{d-1}(-1)^i f_i(P) = 2 - N + \sum_{k=2}^{N-1} C_N^k (-1)^k. \tag{5.3}$$

To calculate the sum on the right side of equation (5.3), we use the expression for the alternating series of combinations (Vilenkin, 1969)

$$C_N^0 - C_N^1 + C_N^2 - \ldots + (-1)^N C_N^N = 0. \tag{5.4}$$

From the series (5.4), taking into account the equalities $C_N^0 = C_N^N = 1, C_N^1 = N$, it follows that the sum $\sum_{k=2}^{N-1} C_N^k (-1)^k$ in the right side of equation (5.3) is N - 2 for N even, and N if N is odd. Therefore, the right side of equation (5.3) coincides with the right side of equation (5.1). This proves that the figures in question are closed convex polytopes and they satisfy the Euler-Poincaré equation.

At the same time, from equation (5.2), since $f_{N+2} = C_N^N = 1$, it follows that the dimension of a cluster of N shells S_j with a common center if at center not atoms is equal $d = N + 2$.

Now consider a cluster of several shells S_j with a common center in which the atom is located. The shell S_j with a central atom has $n_j + 1 = f_0$ vertices, $\frac{n_j t_j}{2} + n_j = f_1$ edges, $\frac{n_j t_j}{2} + j = f_2$ flat faces, $j + 1 = f_3$ – dimensional faces. Substituting the values in the equation (5.1) one can get

$$n_j + 1 - (n_j + \frac{n_j t_j}{2}) + j + \frac{n_j t_j}{2} - (j - 1) = 0 .$$

This proves that the shell S_j with its center has a dimension of 4.

If one continues the edges going from the center of the shell S_j to its vertices and at the appropriate distance arrange more atoms forming the second a shell S_j of a larger size. Such a construction of two the shells S_j with a common atom in center will have $2n_j + 1 = f_0$ $n_j t_j + 2n_j = f_1$ edges, $2j + n_j t_j = f_2$ flat faces, $2j + 2 = f_3$ three-dimensional faces, $f_4 = C_3^2 = 3$ four-dimensional faces. Substituting these values in the equation (5.1), one can get

$$2n_j + 1 - 2n_j(1 + t_j / 2) + 2j + n_j t_j - (2j + 2) + 3 = 2 .$$

This proves that the two shells S_j with common atoms in the center has a dimension 5.

If one continues again the edges coming from the center of the shells S_j, and on these edges one construct the third shell S_j of a still larger size, then such a construction will have $f_0 = 3n_j + 1$ vertices, $f_1 = 3\frac{n_j t_j}{2} + 3n_j$ edges, $f_2 = 3j + 3\frac{n_j t_j}{2}$ flat faces, $f_3 = 3j + 3$ three-dimensional faces,

$f_4 = C_4^2 = 6$ four-dimensional faces, $f_5 = C_4^3 = 4$ five-dimensional faces. Substituting these values into the equation (5.1), one can get

$$3n_j + 1 - 3(n_j + \frac{n_j t_j}{2}) + 3(j + \frac{n_j t_j}{2}) + 3(j+1) + 6 - 4 = 0.$$

This proves that cluster of the tree shells S_j with common atom at center has a dimension 6.

Can write a general expression for the numbers of elements of different dimensions in a cluster with a center for arbitrary number N of shells S_j and give a formula for calculating its dimension. Then in the cluster of N shell S_j with atom in a common center, we have

$$f_0 = Nn_j + 1, f_1 = N\frac{n_j t_j}{2} + Nn_j, f_2 = Nj + N\frac{n_j t_j}{2}, f_3 = Nj + N,$$
$$f_4 = C_{N+1}^2, f_5 = C_{N+1}^3, \ldots, f_{N+2} = C_{N+1}^N. \tag{5.5}$$

Substituting these values into equation (5.1), one can see that in this case the left side of the Euler-Poincaré equation (5.1) takes the form

$$\sum_{i=0}^{d-1} (-1)^i f_i(P) = 1 - N + \sum_{k=2}^{N} C_{N+1}^k (-1)^k. \tag{5.6}$$

To calculate the sum on the right side of equation (5.6), we use the expression (5.4). From the series (5.4), taking into account the equalities $C_N^0 = C_N^N = 1, C_N^1 = N$, it follows that the sum $\sum_{k=2}^{N} C_{N+1}^k (-1)^k$ in the right side of equation (5.6) is $N + 1$ for N even and N -1for N odd. Therefore, the right side of equation (5.6) coincides with the right side of equation (5.1). This proves that the figures in question are closed convex polytopes and they satisfy the Euler-Poincaré equation.

At the same time, from equation (5.5), since $f_{N+3} = C_{N+1}^{N+1} = 1$, it follows that the dimension of a cluster of N shells S_j with an atom in a common center is equal $d = N + 3$.

Q.E.D.

Note that in all considered cases of expansion of the Plato solids, higher-dimensional polytopes of a new type are formed; they do not belong to the known higher-dimensional polytopes of the simplex, cube, cross-polytope types.

From Theorem 5.3, for example, it follows that the dimension of a giant palladium cluster (Vargaftik, et al., 1985), containing 561 atoms in five shells of icosahedrons, is 8. The claims of some authors that the palladium cluster in this case is an E_8 lattice (Shevchenko, 2011) is groundless. The lattice E_8

in an eight-dimensional space with coordinates $(\pm 1, \pm 1, 0, 0, 0, 0, 0, 0)$

units can stand anywhere on the line with arbitrary signs, as well as points with coordinates $(\pm\frac{1}{2}, \pm\frac{1}{2}, \pm\frac{1}{2}, \pm\frac{1}{2}, \pm\frac{1}{2}, \pm\frac{1}{2}, \pm\frac{1}{2}, \pm\frac{1}{2})$

has nothing to do with the structure of a cluster consisting of five icosahedral shells, although it is for such a cluster that the number of atoms gives 561 (Lord, et al., 2006). It is assumed (Coxeter, 1963) that the lattice E_8 corresponds to the polytope of Gosset (Gosset, 1900), which draws from simplexes and cross-polytopes. But from the previous it follows that the polytope corresponding to the giant palladium cluster does not include either simplexes or cross-polytopes.

The outer surface of the clusters of Bergman, Samson, and R-phases (Lord, Mackay, & Ranganathan, 2006) outwardly seems to be different from the icosahedron. However, this difference is not significant. For example, in the Bergman cluster of 45 atoms, the outer surface is supposed to consist of rhombuses. Can note by construction it is not, it is the icosahedron. For example, the clusters from $Mg_{32}(Al, Zn)_{49}$

deform in order to bring it closer to Pauling's three-dimension contohedron (Bergman et al., 1952, 1957). In the Samson cluster of 105 atoms of alloy Mg_6Pd (Samson, 1972), the outer surface is a truncated icosahedron. It is easy to get an icosahedron from this surface, connecting the centers of the pentagons, especially since the initial construction is based on tetrahedrons. The surfaces of the R-phase clusters from

$Mg_{32}(Al, Zn)_{49}, Mo-Cu-Cr, Al_5CuLi_3$
surfaces (Bergman et al., 1952, 1957; Komura et al., 1960; Audier et al., 1988). Therefore, the dimensions of these clusters can be calculated from the formulas obtained in this work, taking into account the number of shells in these clusters. Since the Bergman cluster has two icosahedral shells, according to Theorem 5.3 its dimension is 5. Since the Samson cluster has three shells, its dimension is 6. Since the R-phase cluster has four shells, its dimension is 7.

5.3. Filling the Space with Simplices of Increasing Dimension

As follows from the previous chapter 4, the simplex is often an element of the structure of "inert" substances. The question arises of how the space is filled starting from the polytope of the highest dimension. Here, as well as when filling the space, starting from a three-dimensional figure, an additional edge must emanate from each vertex of the 4-simplex (Figure 5.5.).

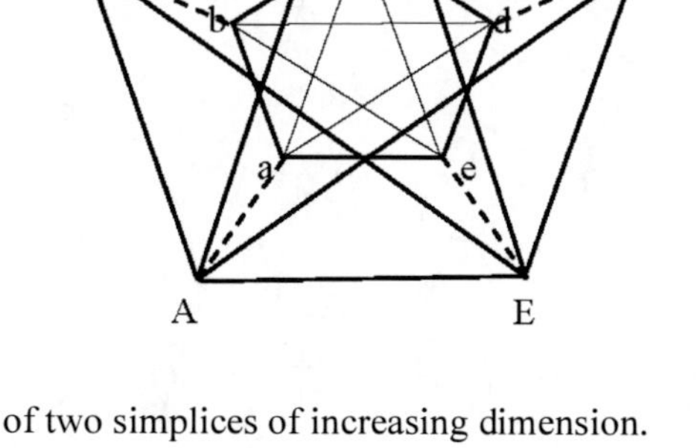

Figure 5.5. Cluster of two simplices of increasing dimension.

Connect the vertices of a larger 4-simplex with edges and count the number of elements of different dimensions in the resulting construction.

The number of vertices on Figure 5.5 is 10 ($f_0 = 10$). The number of edges on Figure 5.5 is 25 ($f_1 = 25$): *ab, bc, cd, de, ea, ac, ad, bd, be, ce, AB, BC, CD, DE, EA, AC, AD, BD, BE, CE, bB, cC, Dd, eE, aA*. The number of

flat faces on Figure 5.5 is 30 ($f_2 = 30$): *abc, acd, ade, abe, bcd, bde,cde, bce, ace, dba, ABC, ACD, ADE, ABE, BCD, BDE, CDE, BCE, ACE, DBA, abAB, bcBC, cdCD, deDE, aeAE, adAD, acAC, bdBD, beBE, ceCE.* The number of free – dimensional faces on Figure 5.5 is 20 ($f_3 = 20$): *abcd, bcde, acde, abde, abce, ABCD, BCDE, ACDE, ABDE, ABCE, BCEbce, CDAcda, DEBdeb, ACEace, ABDabd, ADEade, ABEabe, ABCabc, CBDcbd, DCEdce.* The number of four – dimensional faces on Figure 5.5 is 7 ($f_4 = 7$): *abcde, ABCDE, abdeABDE, abceABCE, abcdABCD, bcdeBCDE, acdeACDE.* Substituting values $f_i, (i = 0,1,2,3,4)$
get

$$10 - 25 + 30 - 20 + 7 = 2.$$

Thus, the dimension of the construction polytope 4-simplex in polytope 4-simplex is equal to 5. It is not a simplex polytope. At the next stage of filling the space, as well as when filling the space with a growing tetrahedron, to fulfill the condition of homogeneity of vertices, it is necessary to connect the vertices of the largest polytope 4-simplex by edges with vertices polytopes 4-a simplex of smaller sizes (Figure 5.6).

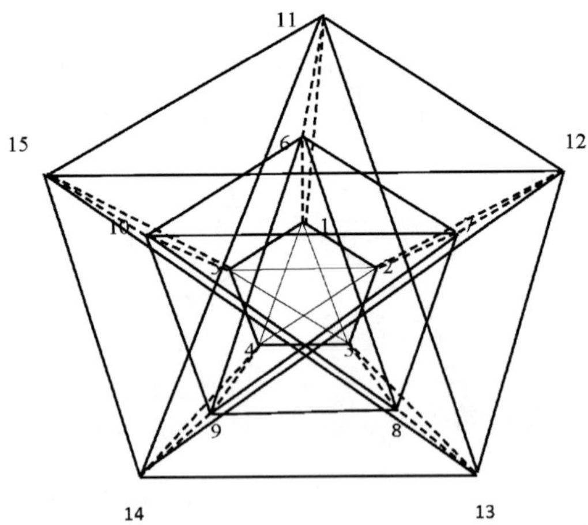

Figure 5.6. Filling the space with polytopes 4-simplex with increasing dimension of the cluster.

Instead of directly counting the number of elements of different dimensions included in the polytope in Figure 5.6, when determining its dimension, you can notice the patterns of their change when moving from one step of filling the space to the next step. Let us denote the number of elements with dimension *i* of the polytope at the step *j* as $f_i[j]$

consider the 4-simplex as the zero step when filling the space with increasing dimension, then we have $f_0[0]=5, f_1[0]=10, f_2[0]=10, f_3[0]=5, f_4[0]=1$. The polytope on the first step of filling has next numbers of elements with different dimension

$$f_0[1]=2f_0[0]=10, f_1[1]=2f_1[0]+f_0[0]=25, f_2[1]=2f_2[0]+f_1[0]=30,$$
$$f_3[1]=2f_3[0]+f_2[0]=20, f_4[1]=2f_4[0]+f_3[0]=7, f_5[1]=1.$$

Let's make sure that the numbers $f_i[1]$ determined by this method completely coincide with the numbers $f_i[1]$ determined by their direct counting according to Figure 5.5.

The polytope on the second step of filling has next numbers of elements with different dimension

$$f_0[2]=3f_0[0]=15, f_1[2]=3f_1[0]+3f_0[0]=45, f_2[2]=3f_2[0]+3f_1[0]+f_0[0]=65,$$
$$f_3[2]=3f_3[0]+2f_2[0]+f_1[0]=45, f_4[2]=3f_4[0]+2f_3[0]=13, f_5[2]=3f_5[1]=3, f_6[2]=1.$$

Substituting values $f_i[2], (i=0,1,2,3,4,5)$

15 – 45 + 65 – 45 + 13 – 3 = 0.

This proves that the polytope constructed as a result of expanding the polytope 4-a simplex in two steps (Figure 5.6) has dimension 6.

5.4. Filling the Space with Cross-Polytopes of Increasing Dimension

When studying the structures of intermetallic alloys, the existence of clusters of tetrahedrons attached to each other, which do not lead to the formation of simplices, was discovered. These clusters are called γ-brass clusters. At first, it was assumed that these clusters were based on a cubic lattice (Bradley, Thewlis, 1926), then they began to consider that the icosahedral is the basis in this cluster (Pauling, 1960). However, these assumptions were not confirmed, and in 2021 it was established (Zhizhin, 2021) that γ-brass clusters are a variant of filling the space of higher dimension with cross-polytopes of increasing dimensions.

5.4.1. Theorem 5.4

The dimension of the polytope (γ-brass cluster) which forms of filling the space with cross – polytope of increasing dimension is $d = 4 + 3n$, where n is the number of cluster shells ($n = 0, 1, 2, \ldots$). The cluster is a d-cross-polytope and the number of elements of the dimension i included in the cluster is determined by the ratio $f_i(d) = 2^{1+i} C_d^{d-1-i}$

5.4.1.1. Proof

Alloys of γ-brass, as well as other intermetallic alloys, are conveniently considered using tetrahedrons (Lord, Mackay, Ranganathan, 2006). Place four atoms at the vertices of the 1234 tetrahedron. Then on each flat face of the 1234 tetrahedron one place another a tetrahedrons (Figure 5.7).

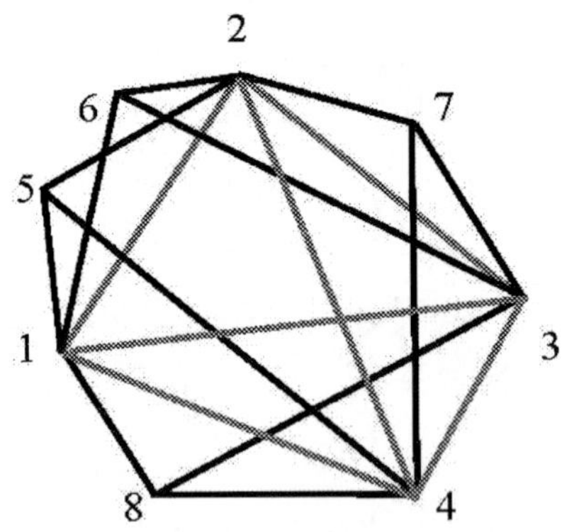

Figure 5.7. The tetrahedron with tetrahedrons on its faces.

Then a figure appears, including 8 atoms (vertices 1-8). Connect the vertices 5, 6, 7, 8 of edges. They also form a tetrahedron. As a result, the resulting figure (Figure 5.8) is a 4-cross-polytope (Zhizhin, 2019 a).

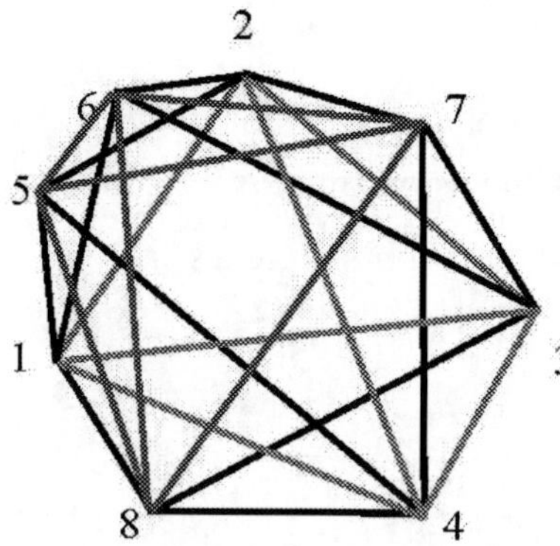

Figure 5.8. The 4-cross-polytope.

In this polytope, in addition to 8 vertices, there are 24 edges, 32 flat triangular faces, and 16 tetrahedrons. Each vertex has no edge connection to some other (opposite) vertex. These unconnected vertices form pairs $\begin{matrix}1&2&3&4\\7&8&5&6\end{matrix}$. Each vertex in the top row does not have a connection with the vertex in the bottom row, just below that vertex. This polytope has dimension 4. Upon further joining of the tetrahedrons to the edges of the original tetrahedron 1234 (two tetrahedrons to the edge), a figure is formed containing 14 vertices. Each newly formed vertex is located opposite one of the edges of the tetrahedron 1234. If we designate the newly formed vertices by a pair of vertices of the corresponding edges of the original tetrahedron 1234, then these are the next vertices (1,2), (1, 3), (1, 4), (2, 3), (2, 4), (3, 4). Among these newly formed vertices, opposite vertices can be distinguished, given the opposite edges of the original 1234 tetrahedron. These opposite vertices also form pairs $\begin{matrix}(2,3)&(1,2)&(1,3)\\(1,4)&(4,3)&(2,4)\end{matrix}$. Connecting the remaining vertices with edges, leaving pairs $\begin{matrix}1&2&3&4\\7&8&5&6\end{matrix}$ unconnected, can get a 7-cross-polytope. In a topologically equivalent form, this polytope of dimension 7 is shown in Figure 5.9.

From the general expression for the number of elements of the dimension i in the d-cross-polytope (Zhizhin, 2013, 2014 c, 2018, 2019 a) $f_i(d) = 2^{1+i} C_d^{d-1-i}$ it follows that the number of vertices in this polytope $f_0 = 2 \cdot C_7^6 = 14$, the number of edges $f_1 = 2^2 \cdot C_7^5 = 84$, the number of triangle faces $f_2 = 8 \cdot C_7^4 = 280$,

the number of tetrahedrons $f_3 = 16 \cdot C_7^3 = 560$,

the number of four-dimension simplexes $f_4 = 32 \cdot C_7^2 = 672$,

the number of five-dimension simplexes $f_5 = 64 \cdot C_7^1 = 448$,

of six-dimension simplexes $f_6 = 128$.

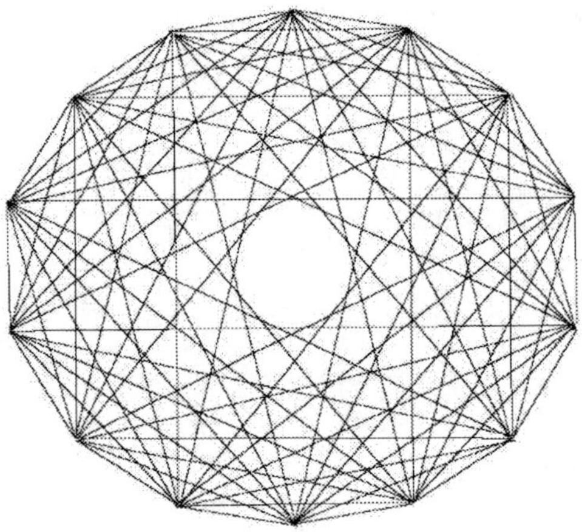

Figure 5.9. The 7-cross-polytope.

Connecting a pair of tetrahedrons to the edges of the tetrahedron 5678, we obtain six more vertices (5, 6), (5.7), (5, 8), (6, 7), (6, 8), (7, 8). Opposite ones can be distinguished among these vertices, considering the opposite of the edges of the tetrahedron 5678 $\begin{matrix}(5,6) & (5,8) & (6,8)\\ (8,7) & (6,7) & (5,7)\end{matrix}$. Connecting the remaining vertices with edges, leaving pairs $\begin{matrix}1\,2\,3\,4\\ 7\,8\,5\,6\end{matrix}$ $\begin{matrix}(2,3) & (1,2) & (1,3)\\ (1,4) & (4,3) & (2,4)\end{matrix}$ $\begin{matrix}(5,6) & (5,8) & (6,8)\\ (8,7) & (6,7) & (5,7)\end{matrix}$ unconnected, we get a 10-cross-polytope. In this polytope the number of vertices is $f_0(10) = 2 \cdot C_{10}^{9} = 20$,

$f_1(10) = 2^2 \cdot C_{10}^8 = 180$,

the number of triangle faces $f_2(10) = 8 \cdot C_{10}^7 = 960$,

the number of tetrahedrons $f_3(10) = 16 \cdot C_{10}^6 = 3360$, the number of four-dimension simplexes $f_4(10) = 32 \cdot C_{10}^5 = 8064$,

the number of five-dimension simplexes $f_5(10) = 64 \cdot C_{10}^4 = 13440$,

the number of six-dimension simplexes $f_6(10) = 128C_{10}^3 = 15360$,

the number of seven-dimension simplexes $f_7(10) = 256C_{10}^2 = 11520$,

the number of either-dimension simplexes $f_8(10) = 512C_{10}^1 = 5120$,

the number of nine-dimension simplexes $f_9(10) = 1024$.

Continuing to attach tetrahedrons to a cluster of 20 vertices, keeping the order of attachment, as in the previous steps, a cluster of 26 atoms can be obtained. This will be the 13-cross-polytope. In this polytope the number of vertices is $f_0(13) = 2 \cdot C_{13}^{12} = 26$, the number of tetrahedrons is $f_3(13) = 2^4 \cdot C_{13}^9 = 45760$

interpenetrating icosahedrons in three-dimensional space, the image of this cluster in the space of dimension 13 in the form of a convex standard cross-polytope can serve. Two such clusters make up an elementary cell of γ-brass.

Thus, it was proved that the addition of tetrahedrons to a γ-brass cluster of 8 atoms, having the form of a 4-cross-polytope, leads to the creation of a number of shells, and the dimension of the cluster when each shell is attached increases on three. The number of elements of different dimensions in a cluster for any shell number *n* (*n* = 0, 1, 2, ...) is determined by the formula established earlier for *d*-cross-polytopes.

Q.E.D.

5.5. Clusters on an Octahedron

Let there be a set of six atoms of osmium bound by a chemical bond (Gubin, 2019). The addition of other osmium atoms to this structure occurs by centralizing the planar triangular faces of the octahedron. As the first stage, atoms attach to triangular faces (located above them on the outside of the octahedron). In this case, faces with attached atoms and faces with non-attached atoms alternate with each other. Since four atoms joined the octahedron, a cluster of ten atoms forms (Figure 5.10).

In order to represent this cluster in the form of a convex figure, it is necessary to supplement it with edges. In this case, for the formation of a convex closed figure, the minimum additional number of edges is 16 (black segments in Figure 5.10). In a topologically equivalent form, this cluster is depicted in Figure 5.11.

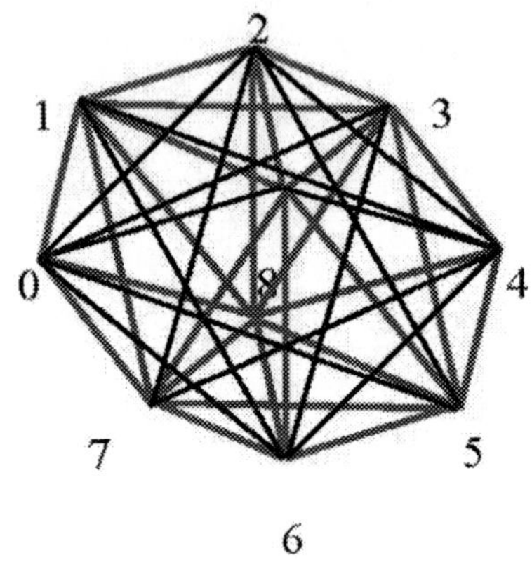

Figure 5.10. The octahedron with attached tetrahedrons on faces of octahedron.

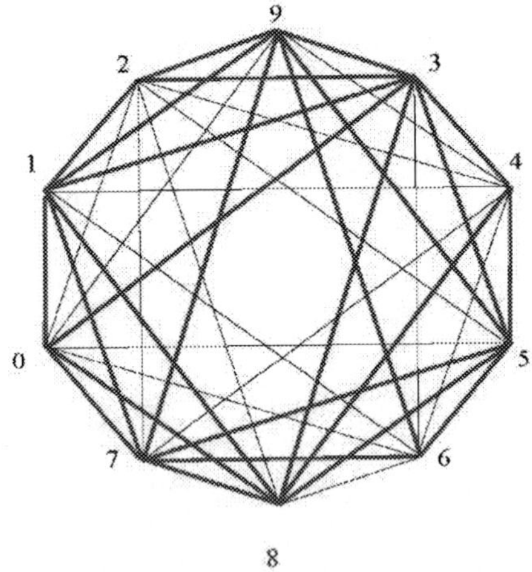

Figure 5.11. The 5 – cross-polytope.

You can see that this is a 5-cross-polytope. Each vertex in Figure 5.11 is connected by an edge to all other vertices with the exception of the opposite vertex. The number of elements of the dimension i in the d-cross-polytope is equal (Zhizhin, 2013, 2014 a, 2018, 2019 b)

$$f_i = 2^{1+i} C_d^{d-1-i} .$$

It have 10 vertices ($f_0(5) = 2^{1+i} C_4^5 = 10$), 40 edges ($f_1(5) = 2^2 C_5^3 = 40$ triangle faces ($f_2(5) = 2^3 C_5^2 = 80$), 80 tetrahedrons ($f_3(5) = 2^4 C_5^1 = 80$ four – dimensional simplexes ($f_4(5) = 2^5 = 32$). The dimension of a figure is determined by the Euler-Poincaré equation (5.1) (Poincare, 1895).

Including the numbers f_i – 40 + 80 – 80 + 32 = 2. This proves that the figure in Figure 5.11 is a convex polytope of dimension 5. It is easy to make sure that with fewer edges the complex of ten atoms will not be convex, since equation (2) will not be satisfied. The next step in the formation of a cluster as a convex figure could be the attachment of atoms to the faces of the octahedron to which the atoms have not yet been attached. There are four such free faces. However, as follows from Figures 5.1 and 5.2, when constructing a convex figure, these free faces turned out to be already occupied tetrahedra without changing the number of vertices. But the edges of these tetrahedrons have only geometric meaning for creating a convex figure. Therefore, attachment to these faces of tetrahedra with chemical bonds is still possible. Then a figure with 14 vertices is formed. Creating a convex figure from it will lead to the formation of a 7-cross-polytope, which has already been shown in Figure 5.9.

Conclusion

After we were convinced in the previous chapter that the so-called "inert" substance creates a complex structure of precious stones, some of which are still beyond the power of a person to describe, this chapter discusses the mechanisms of distribution of the "inert" substance along space. We emphasize that in all cases in these distributions "inert" matter has a higher dimension and therefore has non-Euclidean properties. Geometrically, several such methods of propagation are distinguished, which are characterized by an increase in dimension during propagation with the transition from one step (shell) to another step. This way is a discrete increase of all five bodies of Plato, as an elementary three-dimensional cell in the initial state. Moreover, all the shells are connected with each other by a chemical bond, and in the common center of the shells there may or may not be an atom or molecule of the compound. The obtained mathematical formulas describing these distributions correspond to the formation of a multitude of clusters of substances considered earlier by numerous authors in relation to different compounds. In particular, the structure and dimension of a giant palladium cluster can be easily determined from the obtained formulas. At the same time, it is shown that the frequently mentioned Gosset polytope does not exist, since its existence contradicts the fundamental properties of higher-dimensional polytopes.

The spread of "inert" matter in space can begin not only with the three-dimensional shape of the cell. Since the molecules of "inert" matter, as a rule, have a higher dimension, their distribution in space can begin immediately from a form of higher dimension. Such methods are considered for polytopes 4-simplex and 4-cross-polytope. In particular, the expansion of the 4 – cross -polytope describes the formation of γ-brass clusters, which previously attracted much attention. The distribution of "inert" matter in space also describes the formation of quasi – crystals. In this case, the initial cell is a hyper-rhombohedron with a total dimension of 4 (Zhizhin, 2014 a; Shevchenko, Zhizhin, Mackay, 2013 a, b; Zhizhin, Diudea, 2016; Zhizhin, Khalaj, Diudea, 2016).

Thus, the conquest of space by "inert" matter can occur in different ways, but all of them are characterized by a higher dimensionality of the cells, which increase their dimensionality in the process of propagation.

It should be noted that the spread of "inert" matter in space leads to the formation of new classes of higher-dimensional polytopes. When any of the three-dimensional regular polyhedron (Plato's solids) is increased by one step (for example, tetrahedron or icosahedron), a polyhedron of the same type, but larger, is formed. Moreover, together they form a polytope of dimension 4. This polytope is neither a 4-simplex, nor a 4-cube, nor a 4-cross-polytope. This is a 4 -dimensional polytope of a new type, consisting of two shells connected by a chemical bond. One of the shells can be just a vertex located in the center of the shape. The same situation arises in the expansion of the polytope 4-simplex. The newly formed polytope 4-simplex has a larger size, but together two polytopes 4-simplex, connected by a chemical bond, form a new polytope of dimension 5. It is neither a polytope 5-simplex, nor a polytope 5-cube, nor a polytope 5-cross-polytope. When expanding a 4 – cross-polytope, a polytope of the same type is formed, but the dimension is 7, not 5. There is no increase in the dimension of the polytope by one at each step. However, a cross-polytope of dimension 5 is obtained by attaching tetrahedrons to the triangular faces of an octahedron (expanding the octahedron in a similar way).

Chapter 6

Spatial Models of Sugars and Their Compounds

Abstract

The molecular dimensions of all monosaccharides and their isomers are determined as a function of the number of chiral carbon atoms in the carbon chain. It is shown that in all cases the dimension of molecules is more than three. It was found that the cyclization of monosaccharides leads to an increase in their dimension. Geometric images of monosaccharide molecules in the space of higher dimension and their simplified three-dimensional images are constructed. These images were used to analyze the possibilities for the formation of disaccharides and monosaccharide chains. The results obtained are compared with images of monosaccharides and their chains common in the literature.

Keywords: higher-dimensional space, polytope, sugar molecule, isomer, monosaccharide, chain

Introduction

This and subsequent chapters are devoted to the analysis of the geometry of living matter, its most important biomolecules (sugar, protein, nucleic acids).

Sugar can be considered the basis for the existence of most living organisms since they contain the main content of the energy necessary for their existence. In addition, sugars perform other important functions in organisms. There are three main classes of sugars: monosaccharides, oligosaccharides and polysaccharides. Monosaccharides, or simple sugars, contain only one polyhydroxyaldehyde or polyhydroxyketone structural unit. Their empirical structural formula has the form $(CH_2O)_n$, where $n \geq 3$. Oligosaccharides are made up of short chains of monosaccharides linked by forged bonds. The most common disaccharides are composed of two monosaccharide units. Polysaccharides are long chains made up of hundreds or thousands of monosaccharide units. When depicting sugar molecules, linear forms (Fischer projections) are often used, in which a chain of bonded carbon atoms is represented as a section of a straight line with hydrogen atoms and hydroxyl groups attached to it by single bonds. An oxygen atom is attached to one of the carbon atoms of this chain by a double covalent bond. In solutions, sugar molecules are often present in the form of closed (cyclic) structures. In this case, for their representation, Haworth projections or their modifications are used, in which a closed cycle on a. However, they do not give a spatial picture of the structure of the molecule. Just as in the case of the image of molecules of an "inert" substance, discussed in previous chapters, spatial plane is represented as a "boat" or "armchair" (Metzler, 1980; Lehninger, 1982; Koolman, Roehm, 2013). When depicting the conformations of molecules, spatial three-dimensional images are also used, in which atoms are depicted as balls, and the bonds between them as segments. All these ways of depicting sugar molecules provide certain information about the mutual arrangement of atoms in a molecule images of sugar molecules will be obtained if they look like a polytope (Zhizhin, 2020 a, 2021 b), the dimension of which is determined by the Euler-Poincaré equation (Poincaré, 1895). This chapter is devoted to solving this problem. Sugar molecules containing three carbon atoms in the chain, as well as the most important sugar molecules containing five and six carbon atoms in the chain, are considered. In the functional dimension approximation, the dimensions of all monosaccharides and their isomers are determined as functions of the number of chiral carbon atoms in the carbon chain. Images of polytopes of higher dimension are constructed corresponding to monosaccharides with different carbon chain lengths. Based on the obtained images of higher dimension of monosaccharide molecules, simplified three-dimensional images of them are constructed, which are used to analyze the possibilities for the formation of disaccharides and saccharide chains.

6.1. Spatial Structure of Stereoisomers of Glyceraldehyde and Dihydroxyacetone

The simplest monosaccharides include two trioses (three-carbon sugars): aldose glyceraldehyde (Figure 6.1) and ketose dihydroxyacetone (Figure 6.2). They differ in the position in the carbon chain of the carbon atom with a double bond to the oxygen atom. Each of these monosaccharides can exist in two forms, differing in the different arrangement of hydroxyl groups and atoms of hydrogen relative to the only asymmetric (chiral) carbon atom in this carbon chain. These forms are stereoisomers of the aldose glyceraldehyde and, respectively, the ketose dihydroxyacetone. Figure 6.1 and Figure 6.2 shows stereoisomers *D*-glyceraldehyde and *D*-dihydroxyacetone. Figure 6.3 and Figure 6.4 shows stereoisomers *L*-glyceraldehyde and *L*-dihydroxyacetone.

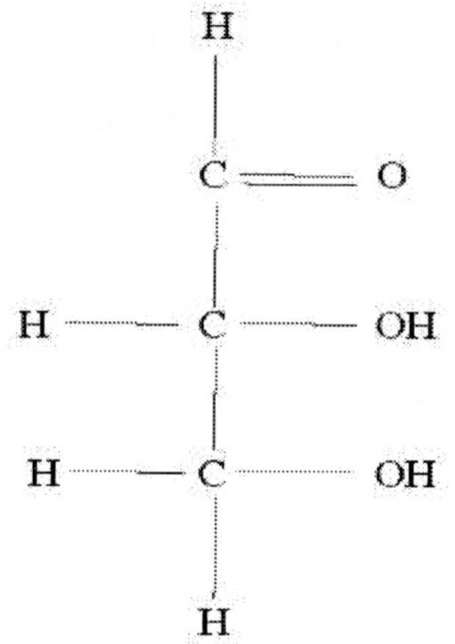

Figure 6.1. D-glyceraldehyde.

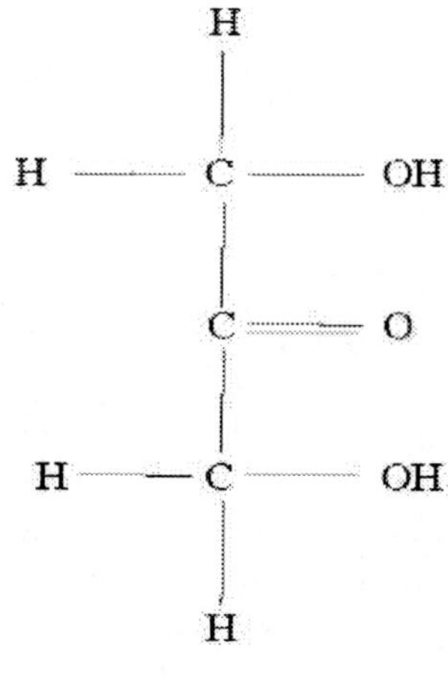

Figure 6.2. D-dihydroxyacetone.

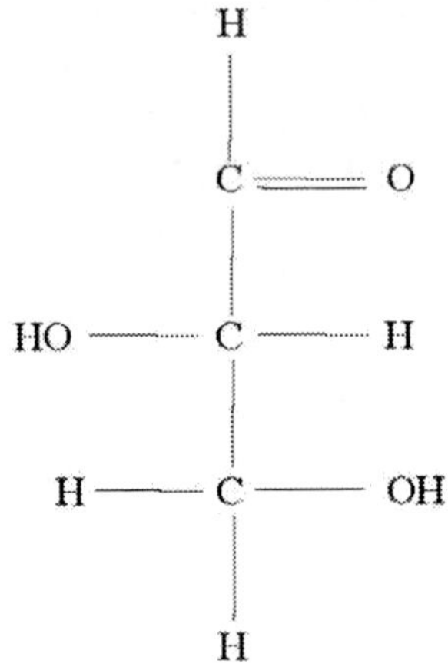

Figure 6.3. L-glyceraldehyde.

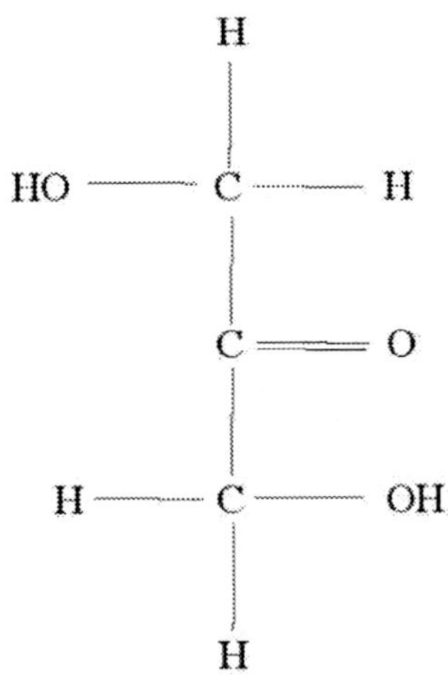

Figure 6.4. L-dihydroxyacetone.

Note that the molecules *D*-glyceraldehyde and *L*-glyceraldehyde contain an aldehyde functional group CHO at the end of the carbon chain. Molecules *D*-dihydroxyacetone and *L*-dihydroxyacetone do not have such a group. The existence of an aldehyde group makes it easier to calculate the dimension of a molecule using the definition of functional dimension (Zhizhin, 2016, 2018, 2019 a).

One will try to calculate the dimension of the molecules *D*-glyceraldehyde and *L*-glyceraldehyde. It follows from Figure 6.1 what if we designate the CHO functional group as the vertex of the desired polytope then it consists of two tetrahedrons with a center. The center of each tetrahedron is the vertex of the other tetrahedron and it is occupied by a carbon atom of the chain.

The corresponding spatial structure is shown in Figure 6.5. In Figure 6.5, the edges corresponding to chemical bonds are indicated by thick solid black lines, and the edges that define only the spatial shape of the polytope are indicated by dotted lines.

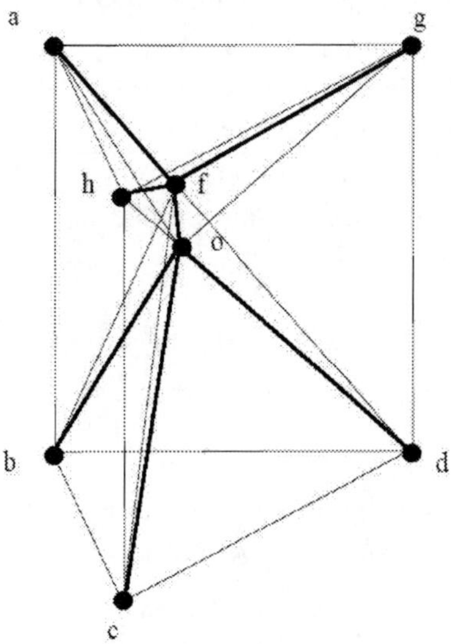

Figure 6.5. Spatial structure of the molecule D-glyceraldehyde.

In this figure there is a tetrahedron *bcdf* with center *o* and a tetrahedron *oahg* with center *f*. Each a tetrahedron with a center is a polytope of dimension 4.

The vertex *f* of the first tetrahedron is the center of the second tetrahedron, and the vertex *o* of the second tetrahedron is the center of the first tetrahedron. Hydrogen (H) atoms are located at the vertices *a, b, c*, hydroxyl groups (OH) are located at the vertices *g, d*, carbon atoms (C) are located at the vertices *o, f.* The functional group CHO is located at the vertex *h*. It is necessary to determine the dimension of the polytope *bcdfahgo*. The polytope in Figure 6.5 has 8 vertices (f_0 *ab, ag, af, ao, ah, gh, gf, go, gd, bf, bo, bd, bc, df, do, dc, hc, hf, ho, cf, co, fo)*. Therefore, f_1 = 22. The polytope in Figure 6.5 has 29 planar faces, of which 26 are the triangles (*aho, afo, ahf, afg, ahg, aog, aob, afb, bfo, bco, bod, bfd, bfc, bcd, ghf, gho, gfo, god, gfd, dfo, dco, dfc, cof, chf, cho, hfo*) and 3 quadrangles (*abdg, hcgd, abhc*). Therefore, f_2

three-dimensional figures, of which 13 are tetrahedrons (*bcfd, ahog, dcfo, bcdo, bfdo, cfdo, ahof, ahgf, hogf, aogf, fgod, fohc, foab*), 6 pyramids (*agbdo, ahbcf, ahbco, agbdf, Chgdf, chgdo*) and one (*ahgbcd)* prism. Therefore, f_3 = 20. It follows from the construction of the polytope in Figure 6.5 that it includes two tetrahedrons with the center *bcdfo* and *oahgf*. In addition to these two polytopes with dimension 4, five 4-polytopes also appear in the polytope in Figure 6.5. Three of these polytopes have as their base three rectangular faces of the prism *ahgbcd*, whose vertices are connected with the vertices *f, o* located inside the prism. To prove their 4-dimensionality, consider one of these polytopes *abhcfo*, since the proofs for the other two polytopes are similar. This polytope has 6 vertices (f_0 = 6); 13 edges (*ab, ah, hc, bc, af, hf, bf, cf, ho, ao, bo, co, fo)*, f_1

faces (*ahf, aho, abo, abf, afo, bfo, boc, ahbc*), f_2

faces (*hfoc, abof, bfoc, afho, ahcbf, ahcbo*), f_3 = 13. Let's now calculate Euler-Poincaré formula for the polytope *P* of dimension *n* (Poincaré, 1895)

$$\sum_{i=0}^{n-1}(-1)^i f_i(P)=1+(-1)^{n-1}. \qquad (6.1)$$

Substituting the obtained values of the numbers of faces of different dimensions into equation (6.1), can find that equation (6.1) is satisfied for $n = 4$

$$6 - 13 + 13 - 6 = 0.$$

It is proved by the 4-dimensionality of the polytope *abhcfo*.

The two polytopes of dimension 4 there are formed by the *ahgbcd* prism with the vertex *f* or *o* inside its. Consider the prism *ahgbcd* with the vertex *f* (the proof for the prism with vertex *o* is similar). The polytope *ahgbcdf* has 7 vertices, f_0 *ah, hg, ag, bd, bc, cd, ab, hc, gd, af, fh, fg, bf, fc, fd*), $f_1 = 15$; 14 two-dimensional faces (*ahg, bdc, ahf, hfg, afg, bfc, fcd, bfd, fhc, afb, fgd, ahbc, hcgd, agbd*), f_2 *ahgbcd, ahgf, bcdf, abdgf, hgcdf, ahbcf*), f_3 numbers of faces of various dimensions obtained for the polytope *ahgbcdf* into equation (6.1) can find that it is satisfied for $n = 4$

$$7 - 15 + 14 - 6 = 0.$$

This proves that the polytope *ahgbcdf* has a dimension of 4.

Thus, for the polytope in Figure 6.5 are $f_0 = 8$, $f_1 = 22$, $f_2 = 29$, $f_3 = 20$, $f_4 = 7$. Substituting these values into equation (6.1) can find that it is satisfied for $n = 5$

$$8 - 22 + 29 - 20 + 7 = 2.$$

This proves that the polytope in Figure 6.5 has dimension 5. Consequently, the molecule of the *D*-glyceraldehyde also has dimension 5.

From Figure 6.3 it follows that in the spatial structure of the molecule *L*-glyceraldehyde, compared to the molecule *D*-glyceraldehyde the arrangement of hydrogen atoms and the hydroxyl group at the vertices *a*, *g* will change. In this structure, a hydrogen atom will be located at the "*g*" vertex, and a hydroxyl group will be located at the "*a*" vertex. Topologically, the structure of the molecule *L*-glyceraldehyde will coincide with the structure of the molecule *D*-glyceraldehyde. Therefore, the dimension of the molecule *L*-glyceraldehyde is also equal to 5.

As follows from Figure 6.2 and Figure 6.4, the stereoisomers of dihydroxyacetones include the CO carbonyl functional group. If we consider the carbonyl group as one of the vertices of the desired polytope, then the spatial structure of the stereoisomers of dihydroxyacetones will include two tetrahedrons with a center having a common vertex of the carbonyl group. The spatial structure of a molecule, for example, *D*-dihydroxyacetone, has the form shown in Figure 6.6.

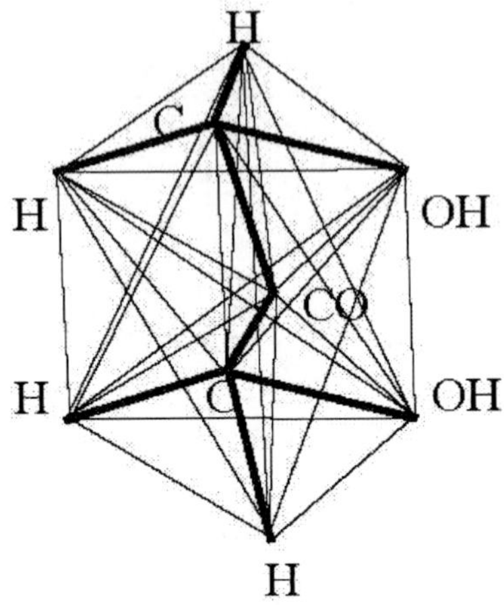

Figure 6.6. The Spatial structure of molecule D-dihydroxyacetone.

In this figure, the edges corresponding to chemical bonds are also indicated by thick solid black lines, and the edges that determine only the spatial image of the molecule are indicated by dotted lines. The polytope in Figure 6.6 has 9 vertices. Moreover, the carbonyl group, as the top of the polytope, incidentally, has 8 edges corresponding to chemical bonds and of edges tetrahedrons. Therefore, for the polytope to be homogeneous, it is necessary that each vertex of the polytope has the same number of edges incident (Zhizhin, 2019 b). It is impossible to reduce the incidence coefficient in this case, since it is determined by the carbonyl group. Therefore, the polytope in Figure 6.6 has the type of a simplex of dimension 8. It is in the simplex that the number of vertices is one more than the dimension of the polytope, since in the simplex each vertex is connected by an edge to all other vertices. Thus, the functional dimension of the *D*-dihydroxyacetone molecule is 8. The number of elements of different dimensions included in the spatial structure of the *D*-dihydroxyacetone molecules is determined by the formulas for the n -dimensional simplex (Zhizhin, 2019 b)

$$f_i(n) = C_{n+1}^{i+1}, i = 0,1,2,...,n. \tag{6.2}$$

According to formula (6.2), the number of elements with dimension zero is 9 ($f_0(8)$), the number of elements with dimension 1 is 36 ($f_1(8)$), the number of elements with dimension 2 is 84 ($f_2(8)$), the number of elements with dimension 3 is 126 ($f_3(8)$), the number of elements with dimension 4 is 126 ($f_4(8)$), the number of elements with dimension 5 is 84 ($f_5(8)$), the number of elements with dimension 6 is 36 ($f_6(8)$, the number of elements with dimension 7 is 9 ($f_7(8)$). Substituting these numbers into the Euler-Poincaré equation (6.1), we see that it is valid for the dimension of the polytope equal $n = 8$

$$9\text{-}36 + 84\text{-}126 + 126\text{-}84 + 36\text{-}9 = 0.$$

This proves that *D*-dihydroxyacetone molecule has dimension 8.

In the case of the *L*-dihydroxyacetone, the dimension of the molecule is also equal to 8, because there is a topological equivalence of the structures of the *D*-and *L*-isomers. The only difference between them is that the hydrogen atom and the hydroxyl group change places in one of the pairs of polytope vertices associated with a carbon atom that is not part of the carbonyl group.

6.2. The Dimension of Linear Molecules of Monosaccharides with a Carbon Length from 4 to 7

If the number of carbon atoms is more than three, then the number of chiral carbon atoms in the carbon chain of the monosaccharide increases (becomes more than one). In this case, each chiral carbon atom gives two isomers with the opposite arrangement of the hydrogen atom and the hydroxyl group relative to the chiral atom. In the aldose family of monosaccharides, the sign *D* is assigned to an isomer if the hydroxyl group at the chiral carbon atom, as far as possible from the carbonyl group, is located on the right. The sign *L* is assigned to an isomer if the hydroxyl group at the chiral carbon atom, as far as possible from the carbonyl group, is located on the left. The total number of aldoses of monosaccharides isomers is 2^m, where m is the number of chiral carbon atoms in the carbon chain. If the number of chiral carbon atoms in the carbon chain of a monosaccharide is greater than one, then when determining the dimension of a linear molecule, as in the case of dihydroxyacetone ketosis, the desired polytope should be sought in the form of a simplex. The dimension of the molecule in this case is equal to the number of vertices of the polytope, reduced by one. In this case, due to the increase in the number of atoms, within the framework of the concepts of functional dimension, both functional groups at the ends of the carbon chain should be considered vertices, i.e., an aldehyde CHO group, and a CH_2 OH group. Then the number of vertices in the polytope with m number of chiral carbon atoms is $3m + 2$. Therefore, the functional dimension d of a linear aldose monosaccharide molecule is $d = 3m + 1$, where m is the number of chiral carbon atoms. Thus, all linear isomers of monosaccharides (aldoses) with three chiral carbon atoms have functional dimension $d = 10$, including such important monosaccharides as *D*-ribose. All linear isomers of monosaccharides (aldoses) with four chiral carbon atoms have functional dimension $d = 13$, including such important monosaccharides as *D*-glucose, *D*-mannose, *D*-galactose. All corresponding polytopes have the form of a higher-dimensional simplex.

In linear ketose isomers, the aldehyde functional group CHO disappears. But now there are CH_2 OH functional groups at both ends of the carbon chain. At the same time, the carbonyl functional group CO remains. Therefore, just as with three chiral carbon atoms, when determining the molecular functional dimension, the carbonyl group, functional groups CH_2 OH, all carbon atoms, hydrogen atoms, and hydroxyl groups should be considered as vertices.

The total number of vertices of the desired polytope in this case is equal to $3m + 3$ (m is the number of chiral atoms). Since the polytope has the simplex type, its functional dimension is one less than the number of vertices, i.e., $d = 3m + 2$.

For example, for isomers of *D*-fructose (three chiral carbon atoms), the number of vertices is 12, and the dimension of the molecule is 11. For isomers of *D*-ribulose (two chiral carbon atoms), the number of vertices is 9, and the dimension of the molecule is 8.

6.3. Functional Dimension of Monosaccharides with a Closed Carbon Chain with Trhee Chiral Carbon Atoms

In monosaccharides, aldoses, the carbonyl group is highly reactive. It attaches to itself groups containing excess electrons, in particular, the OH group closest to the CH_2 OH functional group. In monosaccharides with three chiral carbon atoms, in this case, furanose (five-membered) rings of *D*-

ribose (Figure 6.7), *D*-xylose (Figure 6.8), *L*-arabinose (Figure 6.9) are formed (Metzler, 1980; Lehninger, 1982; Koolman, Roehm, 2013).

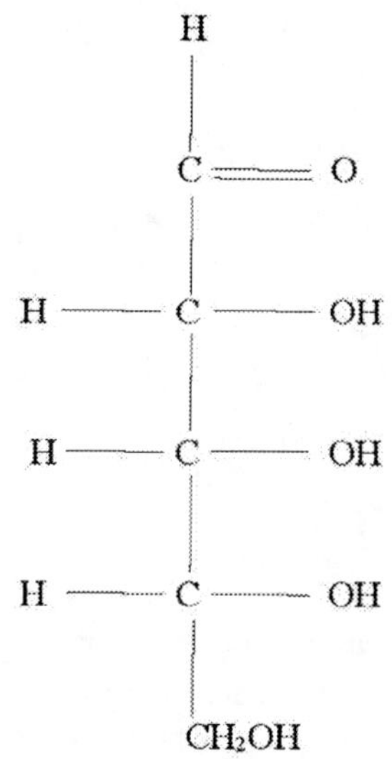

Figure 6.7. The linear molecule *D*-ribose.

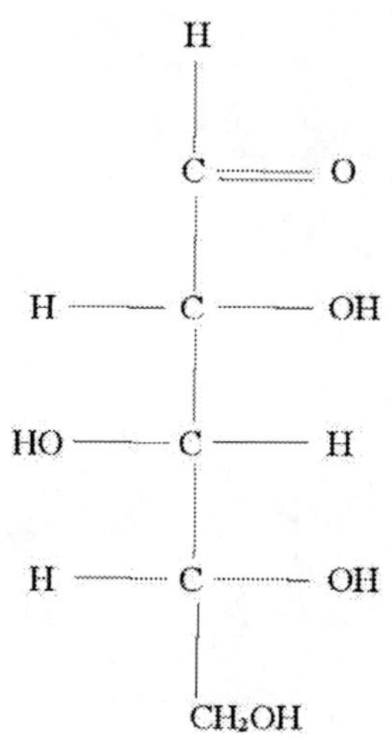

Figure 6.8. The linear molecule *D*-xylose.

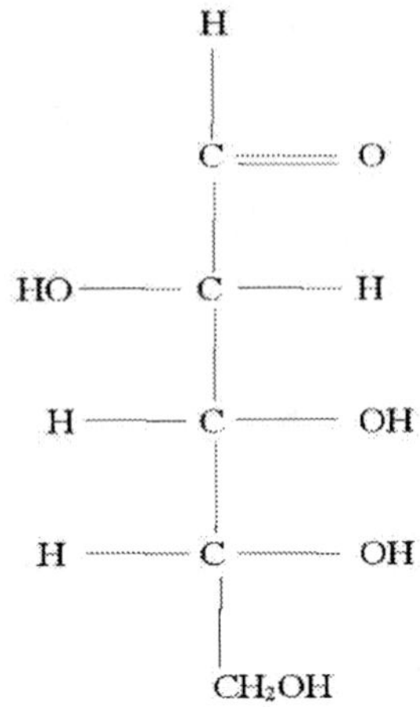

Figure 6.9. The linear molecule *L*-arabinose.

When sugar is cyclized, a new chiral center appears at the carbon atom (anomeric carbon atom) that was previously part of the carbonyl group. The two configurations relative to this atom are designated as the α and β configurations. Taking into account the limitations and sufficient uncertainty of three-dimensional images of Haworth, as well as the higher dimension of linear monosaccharides, we represent furanose rings as polytopes of higher dimension. They are presented respectively in Figure 6.10 (*α-D*-ribose), Figure 6.11 (*β-D*-ribose), Figure 6.12 (*α-D*-xylose), Figure 6.13 (*β-D*-xylose), Figure 6.14 (*α-L*-arabinose), Figure 6.15 (*β-L*-arabinose). For better visualization, we will return to 3D images later, but they will be built from higher dimensional images as some convenient simplifications.

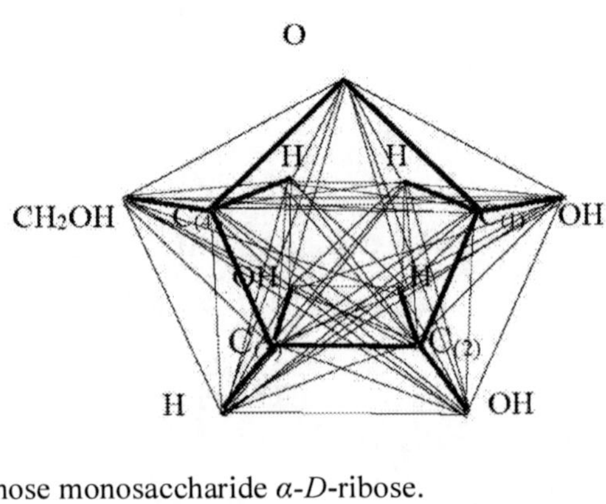

Figure 6.10. Furanose monosaccharide *α-D*-ribose.

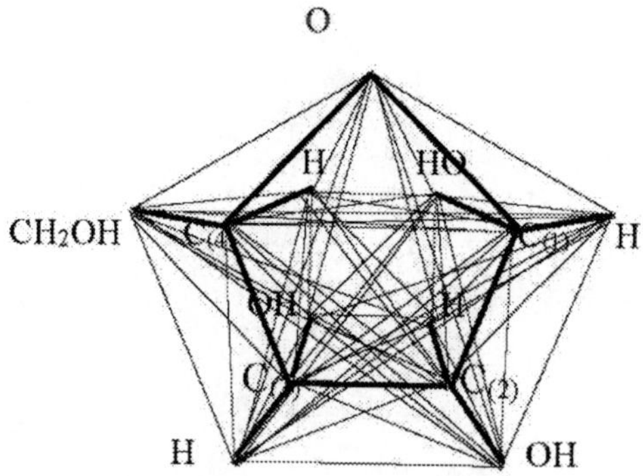

Figure 6.11. Furanose monosaccharide *β-D*-ribose.

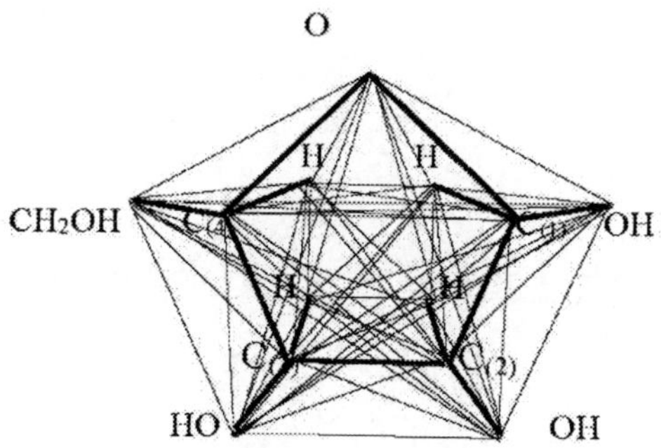

Figure 6.12. Furanose monosaccharide *α-D*-xylose.

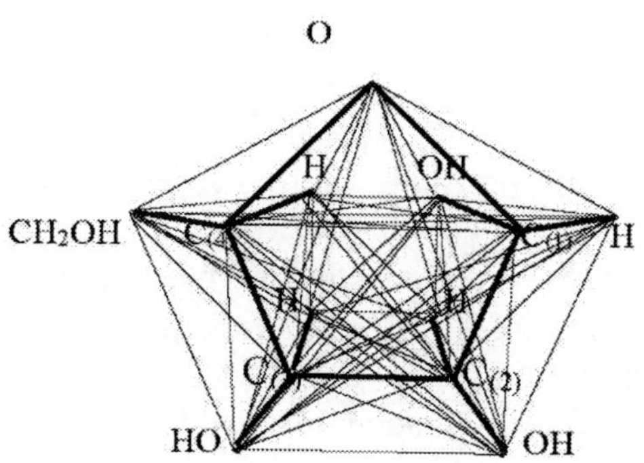

Figure 6.13. Furanose monosaccharide *β-D*-xylose.

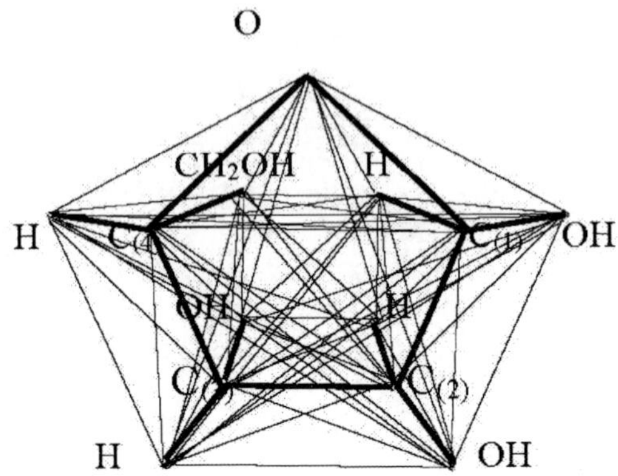

Figure 6.14. Furanose monosaccharide *α-L*-arabinose.

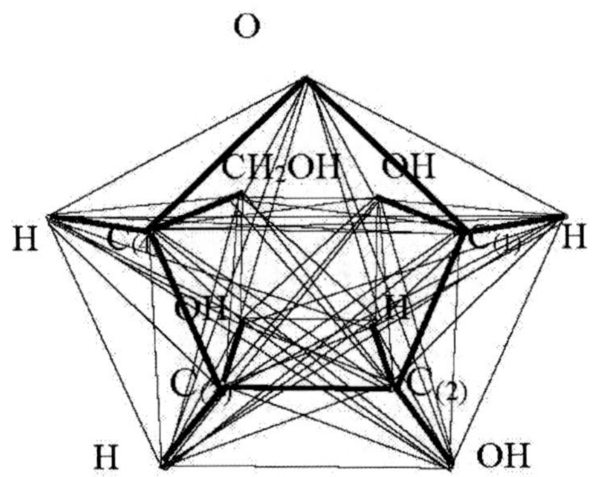

Figure 6.15. Furanose monosaccharide *β-L*-arabinose.

Each of the furanose monosaccharides in Figures 6.10-6.15 can be represented as a polytope. To do this, you need to connect with an edge each vertex in any of these drawings with the rest of the vertices in the corresponding figure. In addition, it is necessary to take into account changes in the arrangement of atoms and functional groups during the transition from linear molecules to molecules with a closed carbon chain. Aldo-pentose *D*-ribose, widely distributed in nature, is part of RNA, DNK and coenzymes of nucleotide nature. In these compounds, ribose is always in the form of furanose. *D*-xylose and *L*-arabinose are part of the polysaccharides of the plant cell wall.

To determine the dimension of furanose polysaccharides (Figures 6.10-6.15), it is necessary to recalculate all the elements of different dimensions that go into polytopes in these figures. For example, the polytope at Figure 6.10 (furanose monosaccharide *α-D*-ribose) has 13 vertices in which there are one oxygen atom, four hydrogen atoms, three hydroxyl groups, four carbon atoms, compound CH_2OH. It is known that the dimension of a simplex is one less than the number of its vertices (Grünbaum, 1967). Therefore, the dimension of the polytope in Figure 6.10, if the Euler-Poincaré equation (6.1) holds for it, equals $n = f_0 - 1 = 12$

simplicial polytope, and all its faces therefore are simplexes. Since all the vertices of a simplex are related to each other, the number of edges is determined by the number of combinations of the number of vertices of two. In this case, the number of edges of the polytope in Figure 6.10 is $f_1 = C_{n+1}^2 = 78$. The number of two-dimensional faces (triangles) of this polytope is $f_2 = C_{n+1}^3 = 286$.

The number of three-dimensional faces (tetrahedrons) of this polytope is $f_3 = C_{n+1}^4 = 715$.

The number of four-dimensional faces (4-simplexes) of this polytope is $f_4 = C_{n+1}^5 = 1287$.

The number of five-dimensional faces (5-simplexes) of this polytope is $f_5 = C_{n+1}^6 = 1716$.

The number of six-dimensional faces (6-simplexes) of this polytope is $f_6 = C_{n+1}^7 = 1716$.

The number of seven-dimensional faces (7-simplexes) of this polytope is $f_7 = C_{n+1}^8 = 1287$. The number of eight-dimensional faces (8-simplexes) of this polytope is $f_8 = C_{n+1}^9 = 715$. The number of nine –dimensional faces (9-simplexes) of this polytope is $f_9 = C_{n+1}^{10} = 286$.

The number of ten-dimensional faces (10-simplexes) of this polytope is $f_{10} = C_{n+1}^{11} = 78$.

The number of eleven-dimensional faces (11-simplexes) of this polytope is $f_{11} = C_{n+1}^{12} = 13$. Substituting the values $f_i, (0 \le i \le 11)$ in the Euler-Poincaré equation (6.1), can see that it holds for n = 12 for polytope on Figure 6.10

$$13\text{-}78 + 286\text{-}715 + 1287\text{-}1716 + 1716\text{-}1287 + 715\text{-}286 + 78\text{-}13 = 0.$$

This finally proves that the polytope *α-D*-ribose has dimension 12 and a simplex type. In Figure 6.10, edges corresponding to valence bonds are indicated by thick solid lines, and edges that have only geometric spatial meaning are indicated by thin solid lines.

Obviously, for each of the furanose monosaccharides on Figures 6.10-6.15, the functional dimension will also be equal to 12 and the polytope will have the simplex type.

Note that as a result of cyclization, the functional dimension of monosaccharide molecules with five carbon atoms increased by two units.

The ketose furanoses *D*-ribulose and *D*-fructose also have a functional dimension of 12. Figure 6.16 shows a closed cycle of monosaccharide *D*-ribulose.

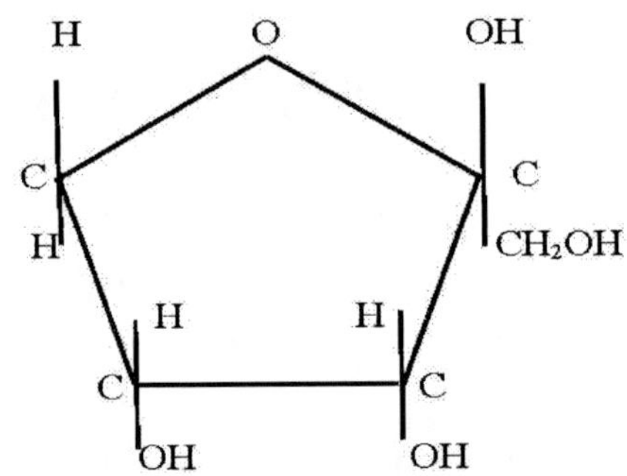

Figure 6.16. Closed cycle monosaccharide D-ribulose.

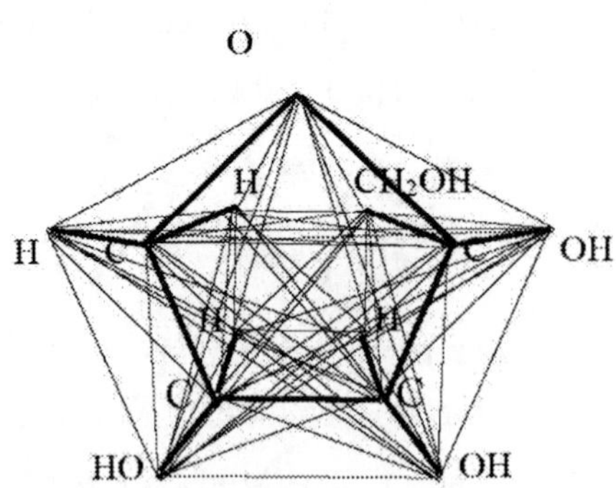

Figure 6.17. Monosaccharide D-ribulose in the form of a polytope.

The same monosaccharide, after joining the edge of each vertex with the remaining vertices, is presented as a polytope in Figure 6.17.

This polytope has 13 vertices. Therefore, its dimension is 12 and it has the type of a simplex.

Figure 6.18 shows a closed cycle of the monosaccharide *D*-fructose.

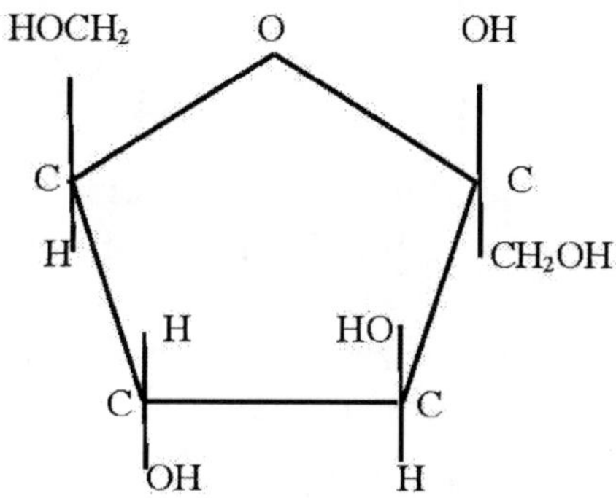

Figure 6.18. Closed-loop monosaccharide D-fructose.

The same monosaccharide, after connecting the edge of each vertex with the remaining vertices, is presented as a polytope in Figure 6.19.

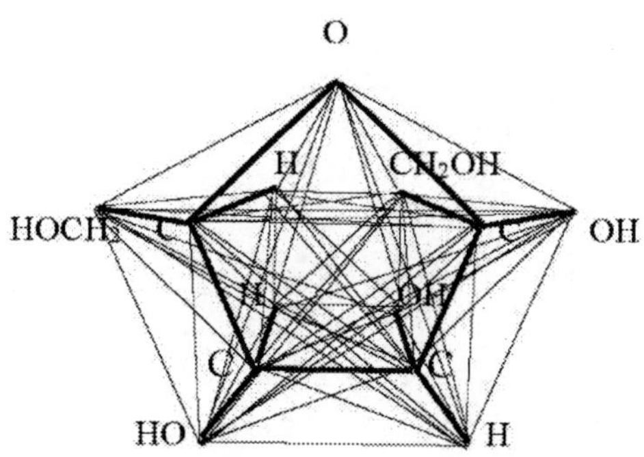

Figure 6.19. Monosaccharide D-fructose in the form of a polytope.

This polytope has 13 vertices, its functional dimension is 12, and it has the simplex type.

It follows from the performed analysis that the cyclization of ketosis also leads to an increase in the functional dimension of the molecule. Moreover, this increases in comparison with aldoses became stronger. The functional dimension of the *D*-ribulose molecule after cyclization increased by 3, the functional dimension of the *D*-fructose molecule after cyclization increased by 4.

D-fructose is found in fruit juices, is part of sucrose and plant polysaccharides.

Ketopentoses *D*-ribuloses are intermediate products of the Ketopentoses *D*-ribuloses are intermediate products of the pentose-phosphate pathway and photosynthesis.

6.4. Functional Dimension of Monosaccharides with a Closed Carbon Chain with Four Chiral Carbon Atoms

In monosaccharides with four chiral carbon atoms, in this case, pyranose (six-membered) rings of *D*-glucose (Figure 6.20), *D*-mannose (Figure 6.21), *D*-galactose (Figure 6.22) are formed.

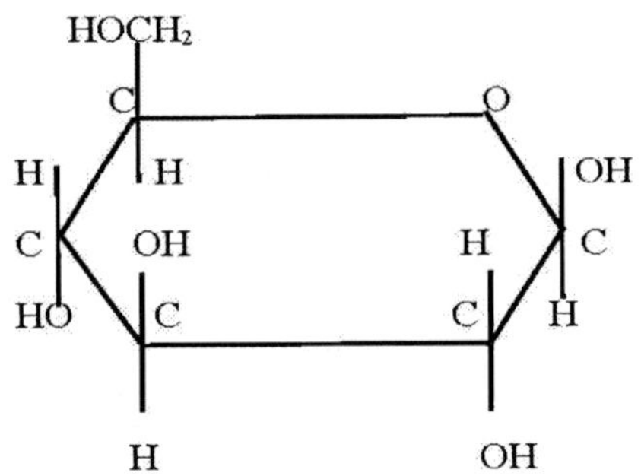

Figure 6.20. The linear molecule D-glucose.

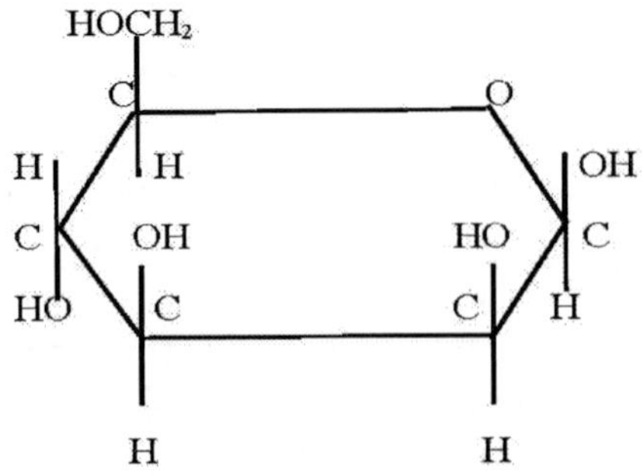

Figure 6.21. The linear molecule D-mannose.

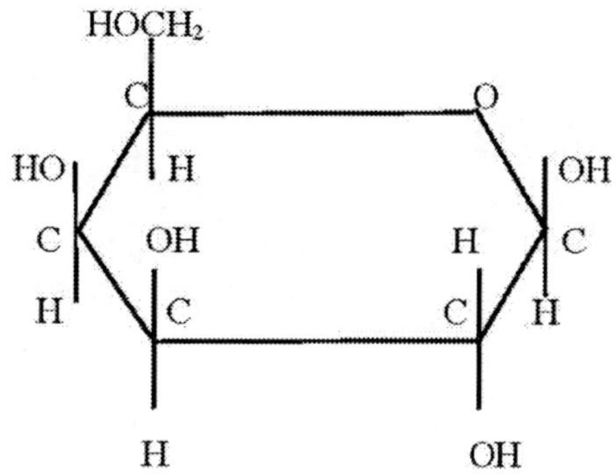

Figure 6.22. The linear molecule D-galactose.

When sugar is cyclized, a new chiral center appears at the carbon atom (anomeric carbon atom) that was previously part of the carbonyl group. The two configurations relative to this atom are designated as the α and β configurations. Taking into account the limitations and sufficient uncertainty of three-dimensional images of Haworth, as well as the higher dimension of linear monosaccharides, we represent pyranose rings as polytopes of higher dimension. They are presented respectively in Figure 6.23 (*α-D*-glucose), Figure 6.24 (*β-D*-glucose), Figure 6.25 (*α-D*-mannose), Figure 6.26 (*β-D*-mannose), Figure 6.27 (*α-D*-galactose), Figure 6.28 (*β-D*-galactose).

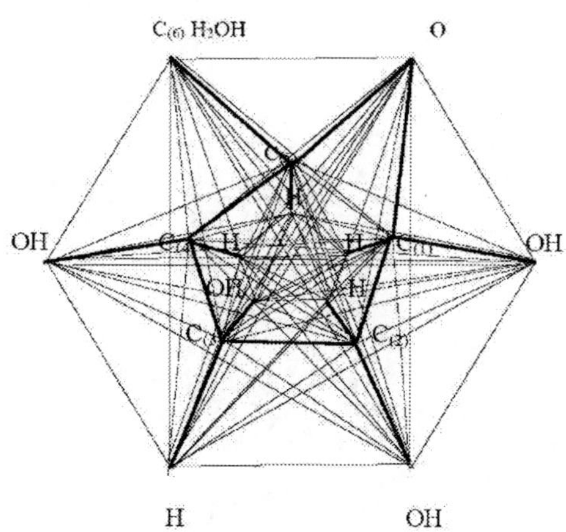

Figure 6.23. Pyranose monosaccharide α-D-glucose.

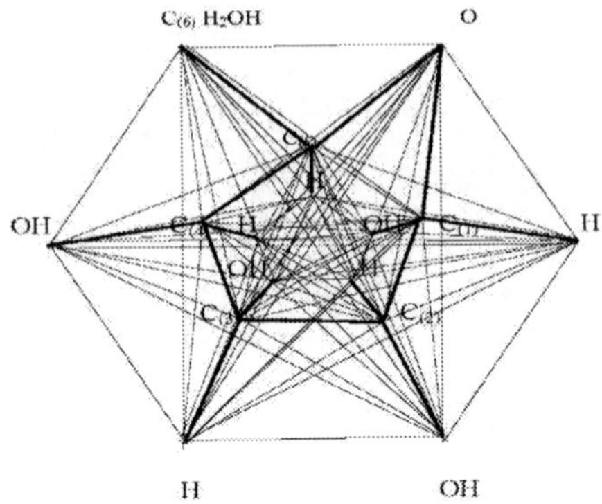

Figure 6.24. Pyranose monosaccharide β-D-glucose.

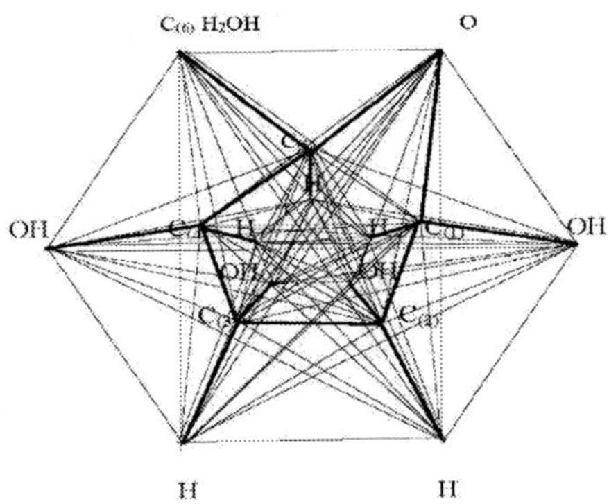

Figure 6.25. Pyranose monosaccharide α-D-mannose.

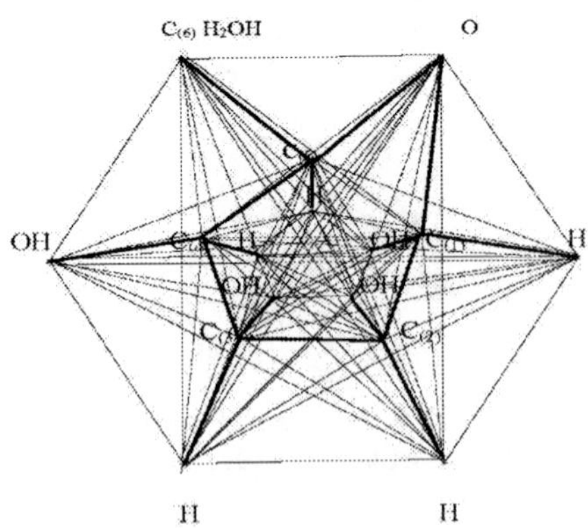

Figure 6.26. Pyranose monosaccharide β-D-mannose.

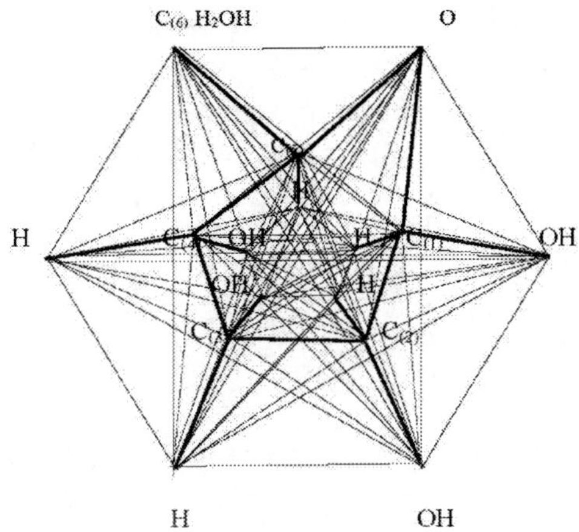

Figure 6.27. Pyranose monosaccharide α-D-galactose.

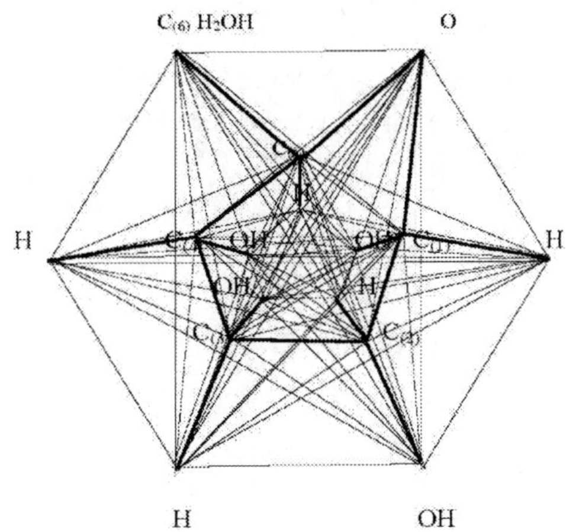

Figure 6.28. Pyranose monosaccharide β-D-galactose.

Each of the pyranose monosaccharides in Figures 6.23-6.28 can be represented as a polytope. To do this, you need to connect with an edge each vertex in any of these drawings with the rest of the vertices in the corresponding figure. In addition, it is necessary to take into account changes in the arrangement of atoms and functional groups during the transition from linear molecules to molecules with a closed carbon chain. The most important aldohexose is *D*-glucose. Glucose polymers, primarily cellulose and starch, make up a significant proportion of all plant biomass. Free form

D-glucose is found in fruit juices, as well as in the blood of humans and animals. The lactose molecule contains *D*-galactose. Along with *D*-mannose, it is part of many glycolipids and glycoproteins.

Edges marked in red in Figures 6.23-6.28 correspond to the chemical bonds in the molecule. The rest of the edges are only geometric sense, as the edges of the polytope.

6.4.1. Theorem 6.1

The molecules of *α-D*-glucose, *β-D*-glucose, *α-D*-mannose, *β-D*-mannose, *α-D*-galactose, *β-D*-galactose are the polytopes type simplex with dimension 15.

6.4.1.1. Proof

To determine the dimension of pyranose polysaccharides (Figures 6.23-6.28), it is necessary for each figure to recalculate all the elements of different dimensions that go into polytopes in these figures. For example, the polytope at Figure 6.23 (pyranose monosaccharide α-D-glucose) has 16 vertices, i.e., $f_0 = 16 = n+1$; 120 edges ($f_1 = C_{n+1}^2 = 120$); 560 triangles ($f_2 = C_{n+1}^3 = 560$); 1820 tetrahedrons ($f_3 = C_{n+1}^4 = 1820$); - simplexes ($f_4 = C_{n+1}^5 = 4368$); -simplexes ($f_5 = C_{n+1}^6 = 8008$); 11440 6-simplexes ($f_6 = C_{n+1}^7 = 11440$); 12870 7-simplexes ($f_7 = C_{n+1}^8 = 12870$); 8-simplexes ($f_8 = C_{n+1}^9 = 11440$); 8008 9-simplexes ($f_9 = C_{n+1}^{10} = 8008$); 10-simplexes ($f_{10} = C_{n+1}^{11} = 4368$); 1820 11-simplexes ($f_{11} = C_{n+1}^{12} = 1820$); 12-simplexes ($f_{12} = C_{n+1}^{13} = 560$); 120 13-simplexes ($f_{13} = C_{n+1}^{14} = 120$); 14-simplexes ($f_{14} = C_{n+1}^{15} = 16$).

Obviously, all isomers of pyranose glucose, including *D*-mannose, *D*-galactose, are topologically equivalent. Therefore, all the values of $f_i, (0 \le i \le 15)$ for these polytopes coincide. Substituting the values $f_i, (0 \le i \le 15)$'s-Poincare equation (6.1) can see that it holds for $n = 15$

$$\sum_{i=0}^{14=n-1} f_i(-1)^i = 2.$$

This confirms that the polytopes in Figures 6.23-6.28 has the dimension $n = 15$.

Theorem 6.1 it is proved.

High dimension of the molecule α-*D*-glucose is due to the fact that it contains a large number of differently oriented electronic atomic orbitals and, consequently, a large amount of energy. This is consistent with the established notions of large energy reserves in glucose, necessary for living organisms. Such an increase in energy and dimension occurs and other saccharides in the formation of closed loops.

Therefore, all of them have dimension 15 and simplex polytope type. Note that in this case, too, cyclization of the molecules led to an increase in its dimension. In a linear form, the dimension of *D*-glucose, *D*-mannose, *D*-galactose was equal to 12.

In particular, the conformation of the β-*D*-glucose are interchanged only a hydroxyl group and a hydrogen atom bound to a carbon atom of the *C* (1) in Figure 6.23. The same conformation occurs in other isomers of glucose. The dimension of the polytope corresponds to the number of vertices of the polytope (not one less than the number of vertices).

6.5. 3D Simplified Image of Pyranose Monosaccharide Molecules

Now, based on the image of the higher dimensional image of the pyranose glucose molecule (Figure 6.23), we can begin to obtain a three-dimensional simplified image of this molecule and then compare this image with known common images. To obtain a three-dimensional simplified model of the molecule α -*D*-glucose, in Figure 6.23 we leave only the edges corresponding to chemical covalent bonds, the edges of the external contour, and the edge connecting the $C_{(1)}, C_{(5)}$

approximation that the ring of carbon atoms $C_{(1)} \div C_{(5)}$

with a side length of 0.15 nm (the standard length of the carbon-carbon chemical bond). Since carbon atoms exhibit tetrahedral coordination, other valence bonds of carbon atoms must be located either above the plane of the ring or below the plane of the ring. It is convenient to assume for further geometric analysis that the atoms of hydrogen and the hydroxyl group located in Figure 6.23 inside the carbon ring are in space above the plane of the carbon ring. While the oxygen atom and the functional groups of the outer contour (larger) are under the plane of the carbon ring. A part of this spatial structure located under the plane of the carbon ring is shown in Figure 6.29.

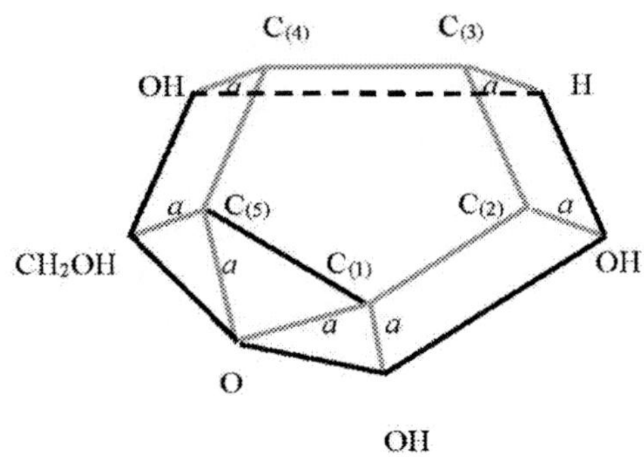

Figure 6.29. The part of the glucose molecule located under the carbon ring.

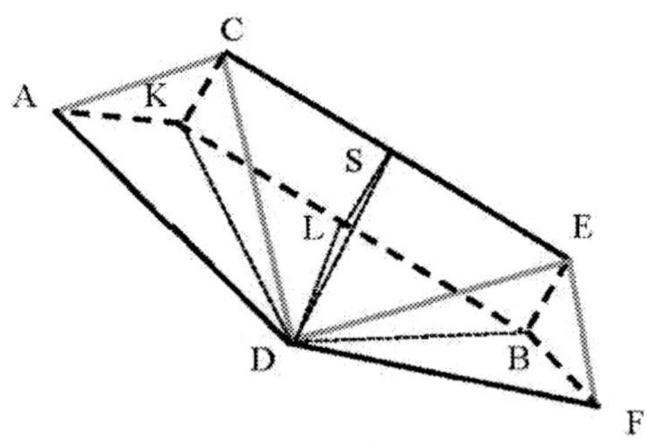

Figure 6.30. The enlarged part of the side surface of the pyramid (Figure 6.29) in the vicinity of the oxygen atom.

The length of the carbon-hydrogen and carbon-oxygen bonds is denoted by the letter $a = 0.1$ *nm*. Then the carbon-carbon bond length takes the form $1.5a$. Thus, the part of the construction of the α -*D*-glucose molecule under the carbon ring is a pyramid with a lower base in the form of an irregular hexagon and an upper base in the form of a regular pentagon.

To determine the yet unknown geometric characteristics of the figure in Figure 6.31, it is first necessary to find the value of the angle of inclination of the side edges of the pyramid to its base. For this, we consider separately the flat lateral faces of the pyramid in the vicinity of the oxygen atom. We double these faces and designate the vertices and some characteristic points of this structure in separate letters (Figure 6.30).

We denote the angle CAK by the letter φ. Given that $\angle AKC = \pi/2$ we have $CK = LS = BE = a\sin\varphi$, so $AC=CD=DE=EF=a$, $AK=a\cos\varphi$. Since the carbon cycle is a regular pentagon, then $\angle AKL = \pi/2 - 54 = 126^o$. Since $CE=1.5a$ and point S is middle of segment CE then $DS = \sqrt{a^2-(3/4)^2a^2} = a\sqrt{7}/4.$ From triangle DSL we have $DL = \sqrt{DS^2-SL^2} = \frac{a}{4}\sqrt{7-16\sin^2\varphi}.$ Therefore, from the triangle KLD we obtain $KD = \sqrt{DK^2+KL^2} = a\cos\varphi = AK.$ Then $\angle KDC = \varphi.$ Besides, $\sin\angle DKL = \frac{DL}{KD} = \frac{\sqrt{7-16\sin^2\varphi}}{4\cos\varphi}.$ Therefore, $\angle AKD = \angle AKL - \angle DKL = 126^o - \arcsin\frac{\sqrt{7-16\sin^2\varphi}}{4\cos\varphi}.$

From the triangle KAD we obtain $\frac{AD}{2a\cos\varphi} = \sin\frac{\angle AKD}{2}$. From here we find

$$AD = 2\cos\varphi\sin(63^o - \frac{1}{2}\arcsin\frac{\sqrt{7-16\sin^2\varphi}}{4\cos\varphi}). \qquad (6.3)$$

Besides, from four-angle $AKLD$ we have $KN/AK=\cos(126^o - 90^o)$. Then, $KN = a\cos\varphi\cos 36^o = LM.AM = AN + NM = a\cos\varphi\sin 36^o + \frac{3}{4}a.$

Therefore, $MD = DL - LM = \frac{a}{4}\sqrt{7-16\sin^2\varphi} - a\cos\varphi\cos 36^o.$

find also one express for segment AD

$$AD = \sqrt{MD^2+AM^2} = \left[(\frac{a}{4}\sqrt{7-16\sin^2\varphi} - a\cos\varphi\cos 36^o)^2 + (a\cos\varphi\sin 36^o + \frac{3}{4}a)^2\right]^{1/2} \qquad (6.4)$$

Comparing (6.3) and (6.4) after transformations, we obtain the equation for finding the angle φ for any a

$$\sin(63^o - \frac{1}{2}\arcsin\frac{\sqrt{7-16\sin^2\varphi}}{4\cos\varphi}) = \left[\frac{1}{2}+\frac{3}{8}\frac{\sin 36^o}{\cos\varphi} - \cos 36^o\frac{\sqrt{7-16\sin^2\varphi}}{8\cos\varphi}\right]^{1/2} \quad (6.5)$$

Solving equation (6.5) numerically, we find that the angle φ = 36º. Thus, the height of the lower part of the α -*D*-glucose molecule is $a\sin\varphi$= 0.058778 nm, the characteristic size of the upper base is 0.243 nm, and the characteristic size of the lower base is 0.243+2$a\cos\varphi$= 0.405 nm.

The angle of inclination of the side edges of the lower part of the model, as follows from the solution, is different from 90. This reflects in averaged form the presence of atoms in the conformations of the α -*D*-glucose molecule located quite close to the equatorial plane of the molecule (Metzler, 1977; Lehninger, 1982; Koolman, Roehm, 2020).

The upper part of the α -*D*-glucose molecule above the carbon cycle should be depicted in the form of a pentagonal prism, directing the volatile bonds from the carbon atoms of the cycle vertically upward. This is consistent with the tetrahedral coordination of atoms around a carbon atom. Thus, the α -*D*-glucose molecule in a simplified three-dimensional form is a pyramid in contact with the carbon cycle with hexagonal and pentagonal bases and a straight pentagonal prism. However, in order to compare the obtained image with the known images of the α -*D*-glucose molecule in the Haworth projection (Koolman & Roehm, 2013), it is necessary to expand so that the pentagonal prism is at the bottom and the arrangement of other atoms and functional groups is consistent. In this case, it is necessary to take into account the possibility of changing the position of the hydroxyl group at the carbon atom $C_{(1)}$

glucose molecule are obtained, shown in Figure 6.31 (β- anomer) and Figure 6.32 (α - anomer).

In the molecules of other pyranoses, the atoms or functional groups partially connected to the carbon atoms of the carbon cycle partially change. In this case, the structure of the molecules, up to functional groups, remains the same for glucose molecules in the pyranose cycle carbon atoms. Therefore, the dimension of molecules such as *D*-mannose, *D*-galactose, *L*-fucose is also equal to 15.

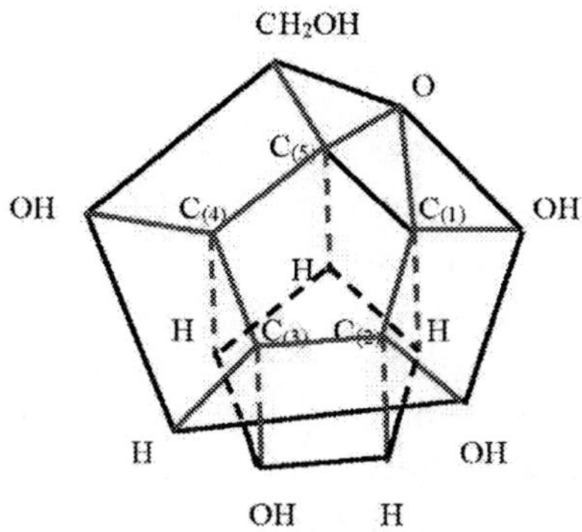

Figure 6.31. The spatial simplified images of the α-D-glucose molecule (β - anomer).

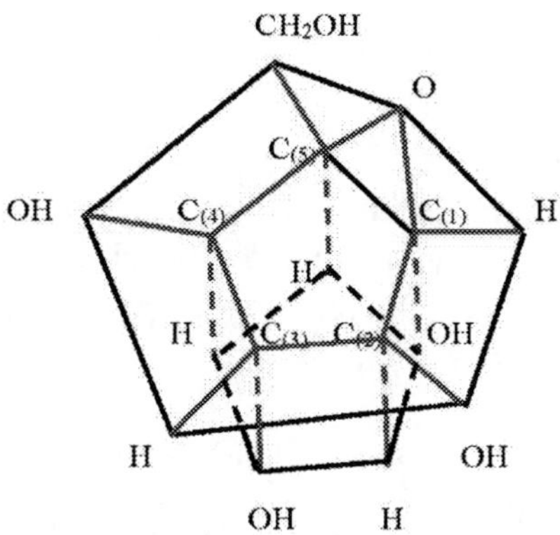

Figure 6.32. The spatial simplified images of the α-D-glucose molecule (α - anomer).

Thus, we see that a simplified three-dimensional image of a pyranose monosaccharide molecule has the form of two three-dimensional polytopes stacked on top of each other. The lower figure is cylindrical prism with a regular pentagon base. The top figure is a pyramid with pentagonal and hexagonal bases. These figures are connected by a carbon cycle. We emphasize that this image was obtained as a result of the analysis of the image in the space of higher dimension and numerical research. The resulting image does not in any way resemble an armchair, which is widely used in the literature as an image of a pyranose monosaccharide molecule.

6.6. Monosaccharide Chains

Representations of the saccharide molecules in the form of polytope simplifies the understanding of the formation of polysaccharides.

Monosaccharide molecules combine with each other thanks to the combination of two hydroxyl groups with the release of a water molecule. The remaining oxygen atom connects the remaining part of the monosaccharides. The most common chains of *α -D*-glucose monosaccharide residues. Here, residual *α-D*-glucose molecules can be joined through an oxygen atom, connecting carbon atoms 1-1, 1-4, 4-4, 4-1. In some cases, the carbon atom $C_{(2)}$ in the cycle will be involved in the compound. Branching chains of glucose molecules occurs at the sixth carbon atom in the functional group CH_2OH. Monosaccharide chains have different forms depending on the type of these molecules, i.e., depending on whether these molecules are β- anomers or *α*-anomers. Using the obtained simplified three-dimensional models of *α-D*-glucose molecules, we consider in more detail these compounds in a metric image, taking into account the angles between the valence bonds. Let two β- anomers of the *α-D*-glucose molecule join together, linking the carbon atom $C_{(1)}$ of the first *α-D*-molecule to the carbon atom $C_{(4)}$ of the second *α -D*-molecule (lactose). Based on the geometric image in Figure 6.33, you can imagine a top view of this connection (Figure 6.33).

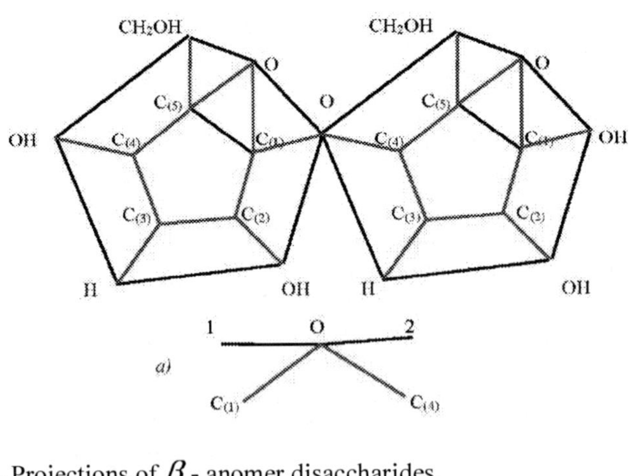

Figure 6.33. Projections of β- anomer disaccharides.

It is easy to see that the same attachment of the third *α-D*-glucose molecule leads to a linear arrangement of molecules when viewed from above on this compound. Linearity will not change with the following similar connections. However, given the three-dimensionality of the image, you should look at this connection from a different view. Figure 6.33a) shows an image of the junction of two *α-D*-glucose molecules when viewed from the front. The black segments in Figure 6.33 a) are the traces of the intersection of the upper bases of the prism with a plane passing through the valence bonds of the oxygen atom and carbon atoms. According to the obtained solution of equation (6.4), the angle between the base of the prism and the valence bond is 36 degrees. Both the first *α -D*-glucose molecule and the second *α -D*-glucose molecule. The valence angle at the oxygen atom in compounds with two carbon atoms in the chain of *α -D*-glucose molecules, as you know, is slightly larger than the normal valence angle and is about 112 degrees. Thus, the angle 1O2 in Figure 6.33 a) is 176 degrees. This means that the base of the second *α -D*-glucose molecule has a slope to the base of the first *α-D*-glucose molecule other than zero. More precisely, this slope is 4 degrees, therefore the second *α-D*-glucose molecule, when viewed in full view, turns to the left relative to the first *α-D*-glucose molecule. It is clear that such an addition of a third *α-D*-glucose molecule will increase the angle of inclination of this molecule relative to the first molecule. Thus, the statement about the linearity of the chain of *α-D*-glucose molecules when connecting them in the case of anomers is not entirely accurate. One can say that this statement is somewhat one-sided.

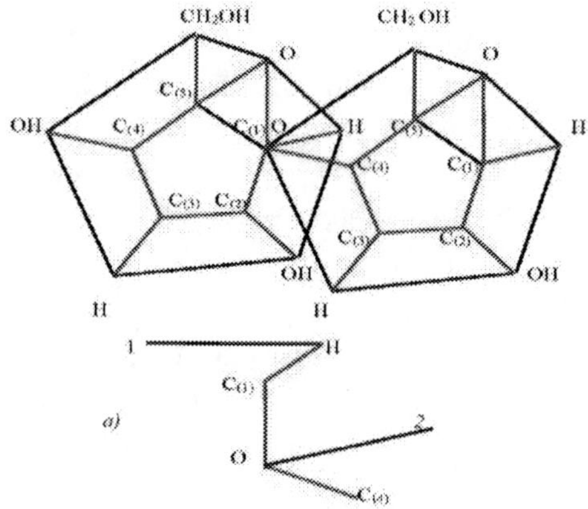

Figure 6.34. Projections of α-anomer disaccharides.

Let two α-anomers of the α-*D*-glucose molecule join together, linking the carbon atom $C_{(1)}$ of the first α -*D*-molecule to the carbon atom $C_{(4)}$ of the second α-*D*-molecule (maltose). Based on the geometric image in Figure 6.32, you can imagine a top view of this connection (Figure 6.34).

It is easy to see that the same attachment of the third α-*D*-glucose molecule leads to a nonlinear arrangement of molecules when viewed from above on this compound. Nonlinearity will save with the following similar connections. Thus, the chain of molecules turns to the right. However, given the three-dimensionality of the image, you should look at this connection from a different view. Figure 6.34 a) shows an image of the junction of two α-*D*-glucose molecules when viewed from the front. The black segments in Figure 6.34 a) are the traces of the intersection of the upper bases of the prism with a plane passing through the valence bonds of the oxygen atom and carbon atoms. According to the obtained solution of equation (6.4), the angle between the base of the prism and the valence bond is 36 degrees both the first α-*D*-glucose molecule and the second α-*D*-glucose molecule. Thus, on the Figure 6.34 a the angle 1H $C_{(1)}$ is 36 degrees and the angle $C_{(1)}OC_{(4)}$ is 112 degrees. Since, the angle $C_{(4)}O2$ is 36 degrees too, so the angle between bases 1H and O2 is 14 degrees. This means that the base of the second α-*D*-glucose molecule has a consequently slope to the base of the first α-*D*-glucose molecule to the left. It is clear that such an addition of a third α-*D*-glucose molecule will increase the angle of inclination of this molecule relative to the first molecule. In the chains of β- anomers, one of the projections is linearly transmitted along the chain, and the projection perpendicular to it along the chain along a curved line. In chains of α-anomers (spirals), it is observed that when moving along a chain of projections in mutually perpendicular planes, the pyranose molecules rotate in opposite directions (Figure 6.35).

Thus, chains of α-anomers of α-*D*-glucose molecules rotate simultaneously in opposite directions in perpendicular planes.

The study of furanose chains using simplified three-dimensional models was carried out using nucleic acids as an example, in which furanose models were connected to each other using phosphoric acid residues. These studies showed that, when nucleic acids bound by nitrogen bases interact, there is a latent degree of freedom of arrangement of nitrogen bases in higher dimensional space, leading to an unambiguous corresponding number of nucleotides taking into account their possible location in space and the number of amino acids (Zhizhin, 2020, b, c; 2021 a).

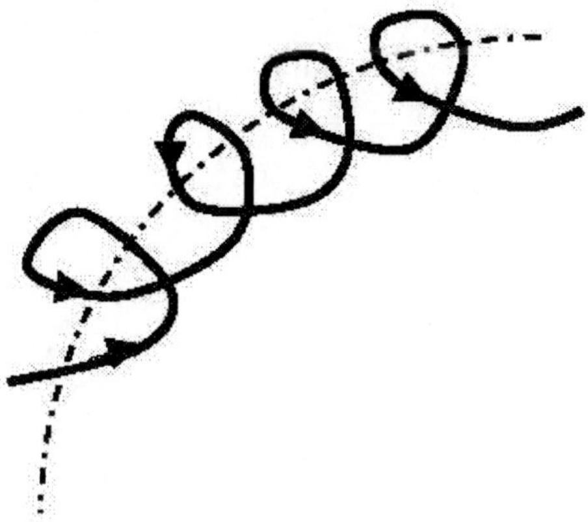

Figure 6.35. The chain of α-anomers (spirals).

Conclusion

As part of the definition of functional dimension, in which all functional groups of monosaccharide molecules are taken as vertices of the corresponding geometric figure, analytical expressions are obtained for calculating the dimension of polytopes of monosaccharide molecules depending on the number of chiral carbon atoms. They refer to both linear molecules and molecules with a closed carbon cycle (furanoses and pyranoses). It has been shown that the cyclization of monosaccharides leads to an increase in their dimension. The geometric shape of monosaccharide molecules has the form of a polytope of the simplex type of higher dimension. Based on the type of polytope of higher dimension corresponding to the pyranose monosaccharide molecule and calculations, a three-dimensional approximate image of the pyranose monosaccharide molecule was obtained. It has the form of a pentagonal prism connected by a carbon cycle to a pyramid, which has a pentagonal and hexagonal base. This image is fundamentally different from the commonly used in the literature image of the pyranose molecule in the form of an "armchair." The resulting simplified three-dimensional image of the pyranose monosaccharide molecule was used to analyze the possibility of the formation of disaccharides and monosaccharide chains.

It has been found that the statement about the linearity of the connection of β-anomers of monosaccharides, widespread in the literature, is not correct since a deviation from linearity is observed in the chain of monosaccharides.

An even greater deviation from linearity in the monosaccharide chain is observed when α-anomers are combined. Here, the possibility of the formation of spirals with simultaneous rotation in opposite directions was discovered.

Chapter 7

The Theory of the Folder and Native Structures of the Proteins

Abstract

Spatial models of the β-structures of protein molecules, forming layers of amino acids, in principle, of unlimited length for different conformations have been constructed. It is shown that the simplified flat Pauling models do not reflect the spatial structure of these layers. Using the recently developed theory of higher-dimensional polytopic prismahedrons, models of the volumetric filling of space with amino acid molecules are constructed. The constructed models for the first time mathematically describe the native structures of globular proteins. A model that simultaneously contains β-structures in parallel and anti-parallel positions and α-spirals. The turn of the polypeptide chain, which is often found in globular proteins, has been studied in the higher dimensional space.

Keywords: polypeptide chain, amido acid, polytope, cross-polytope, amyloid, protein

Introduction

Each organism contains thousands of proteins that perform various functions. Structural proteins are responsible for maintaining the shape and structure of cells and tissues. Transport bands ensure the transfer of oxygen and carbon dioxide, ions and metabolites. Regulatory proteins are responsible for signaling. The largest group of proteins includes enzymes that control all processes in the body associated with chemical reactions (Dixon, Webb, 1979). In their form, proteins can represent a linear chain of amino acids linked by a peptide bond (primary structure). The secondary structure includes sections of the peptide bond with a certain conformation, stabilized by hydrogen bonds. The conformation of a protein, consisting of elements of the secondary structure and unstructured fragments, is called the tertiary structure. Due to the absence of non-covalent interactions, many proteins form symmetrical complexes with each other, which are called quaternary structures (Metzler, 1977; Lehnindger, 1982; Koolman, Roehm, 2013). The most common variants of the secondary structure of proteins are α-helices and β-layers. With such conformations, hydrogen bonds arise between adjacent helices and between adjacent layers. If the peptide chains in the β-layer go in the same direction, then the β-layers are called parallel. If the peptide chains in the β-layer run in opposite directions, then the β-layer is called antiparallel. Stacked in even layers, antiparallel β-layers form a conformation called fibroin. Soluble proteins differ from insoluble (fibrillar) proteins by the formation of spherical globules. Soluble proteins differ from insoluble (fibrillar) proteins by the formation of spherical globules. In a biologically active state, globular proteins are characterized by a specific structure-a native conformation. This conformation simultaneously includes both α-helices and β-layers. The linear Pauling model is currently used to describe various protein conformations (Pauling, Corey, 1950, 1953; Pauling, et al., 1951; Corey, Pauling, 1953). However, it has recently (Zhizhin, 2020 a, 2021 a, 2022 a) been shown that the Pauling model does not reflect and even contradicts the spatial structure of amino acid polypeptide molecules without taking into account their higher dimension. At present, the question of the correct description of protein conformations is associated, in particular, with the need to find ways to combat so far incurable prion diseases Alzheimer`s, Parkinson`s, Kreietzfeld-Jakob`s and others. In all these diseases, insoluble large protein aggregates with β-structure are formed (β-amyloids), disrupting the brain, leading to neurodegeneration and disrupting protein metabolism. The structure of these formations is still unknown, and ways to combat their formation also remain unknown. It has been established that amyloids can be formed not only in the brain, but also in the liver, and enter the brain with lipoproteins. In this chapter, protein of different conformations in the form of fibroin and native structure are considered taking into account their real extent in the higher-dimensional space.

7.1. Dimensions of Protein Molecules

Monomer units which are built of proteins are the 20 standard amino acids. These small molecules containing two different chemical functional groups capable of reacting with each other to form a covalent bond. Amino acids are chemical compounds that simultaneously include a carboxyl group (-COOH) and an amine group (-NH_2). These groups are covalently linked to a carbon atom (α-carbon), to which a hydrogen atom H and a side chain R are also covalently attached (Koolman & Roehm, 2013). Connection which determines the formation of protein polymer is called a peptide bond. In the formation of such a connection by joining together-COOH and -NH_2 with secretion a molecule of water. Amino acids have two enantiomorphic forms L and D, which are Fischer planar views shown in Figure 7.1. From Figure 7.1 it can be seen that the two enantiomorphic forms are related by mirror image relative to the line passing through the amine and carboxyl groups.

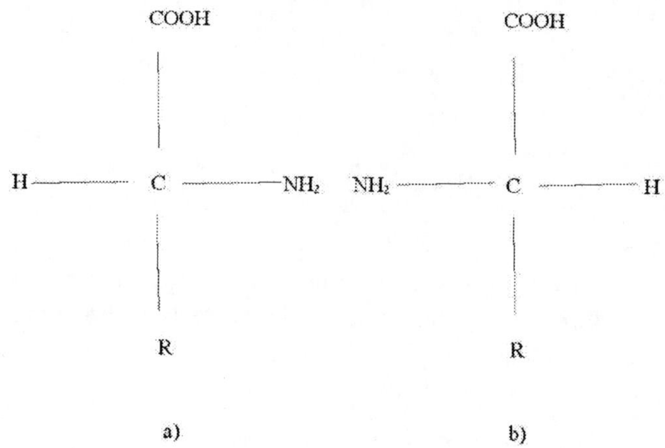

Figure 7.1. Enantiomorphic forms of the amino acid molecule.
a) L-amino acid, b) D-amino acid.

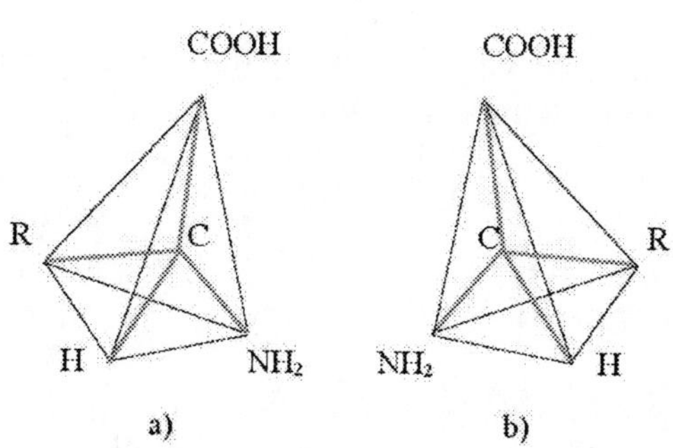

Figure 7.2. Spatial images of amino acids. a) L-amino acid, b) D-amino acid.

The side chains in Figure 7.1 have 20 variants in classic amino acids (Koolman, Roehm, 2013). They determine the properties of the corresponding amino acids. Fisher's projections do not reflect the spatial structure of amino acids, although it is clear that these are spatial objects. Spatial images of enantiomorphic amino acids (Zhizhin, 2016 a) are presented in Figure 7.2.

In Figure 7.2, covalent bonds are indicated in red. The rest of the segments define the spatial shape of the molecule, they are marked in black.

From Figure 7.2 it follows that the amino acid molecule is a tetrahedron with a center. The dimension of such a polyhedron can be determined by the Euler-Poincaré equation (Poincaré, 1895)

$$\sum_{i=0}^{n-1}(-1)^i f_i(P)=1+(-1)^{n-1}. \qquad (7.1)$$

There n is dimension of a polytope P, $f_i(P)$

with dimension i in the polytope P.

From Figure 7.2 it follows that in any of the two enantiomorphic forms of the amino acid molecule, the number of vertices is 5, the number of edges is 10, the number of two-dimensional faces is 10, the number of three-dimensional faces is 5. In this case $f_0=5, f_1=10, f_2=10, f_3=5$.

Substituting the obtained values $f_i, i=0\div 3$

obtain,

5 – 10 + 10-5 = 0.

Thus, Euler-Poincaré's equation for a tetrahedron with a center holds for $n = 4$. This proves that a tetrahedron with a center (amino acid molecule) has dimension 4.

7.2. Linear Polypeptide Chain Structure

The study of diffraction patterns of crystals of amino acids (Pauling, Corey & Branson, 1951; Pauling & Corey, 1950, 1953; Corey & Pauling, 1953) made it possible to establish some values of bond bonds and bond angles in an amino acid. However, the interpretation of the spatial arrangement of atoms in an amino acid requires additional instructions. An example of the possibility of ambiguous interpretation of diffraction patterns can be found in numerous works on the description of quasicrystals. In particular, the apparent icosahedrical symmetry of the diffraction patterns of quasicrystals was proposed to be explained (Pauling, 1987) by multiple twinning of cubic crystals. Pauling based his arguments on the radial arrangement of the spots in the diffractograms. However, high-resolution photomicrographs refuted this assumption (Gratias, Cahn, 1986). The statement about the absence of translational symmetry in quasicrystals (Shechtman, et al., 1984) was constructed using the representations of the three-dimensional geometry of Euclid. Later this statement was refuted (Zhizhin, 2013 a, b; Zhizhin, 2014; Zhizhin, Diudea, 2016). It was proved that translational symmetry in the diffractograms of quasicrystals appears if the diffractograms are considered as projections of structures from the space of higher dimensions. When studying diffraction patterns of amino acid crystals in the works of Pauling joint authors the values of a number of parameters were not determined, for example, bond lengths *R-C* depending on the species *R*, bond angles between carbon atoms and nitrogen atoms, between hydrogen atoms and *R*, etc. To compensate for these disadvantages, it was proposed to use a simplifying assumption about the arrangement of a group of atoms of a peptide bond and centers of tetrahedrons in a plane (Zhizhin, 2020 a) (Figure 7.3).

Figure 7.3. A simplifying assumption about the arrangement of a group of atoms of a peptide bond.

Under this assumption, knowledge of the indicated quantities was not required. It should be noted that in subsequent publications by other authors (Metzler, 1977; Lehninger, 1982; Koolman, Roehm, 2013) with a reference to Pauling's work, it was argued that the arrangement of these atoms in the plane was proved in Pauling's works. But this is not so, to be convinced of this it is enough to refer to these works. This unproven statement became the basis for the construction of three-dimensional models of protein molecules of various conformations in Pauling's works. Let us consider this issue in detail when constructing spatial structures of a linear polypeptide chain. In this case, the atoms of amino acid molecules must be located on a set of parallel lines. Then the polypeptide chain of two molecules of *L*-amino acids, taking into account the four-dimensionality of amino acid molecules, has the form shown in Figure 7.4.

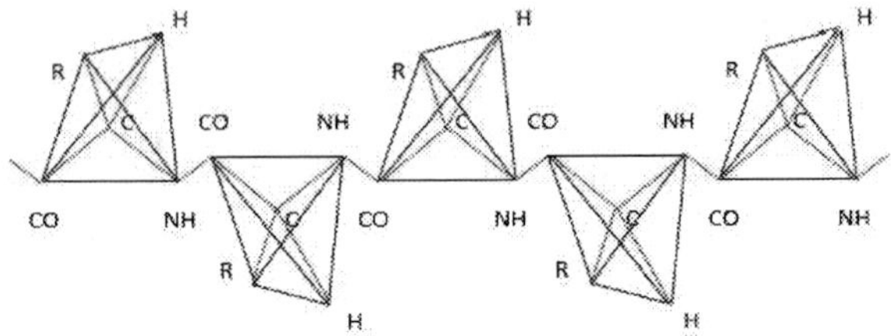

Figure 7.4. Linear polypeptide chain of L-molecules amino acids.

In the case of a *D*-amino acid, such a polypeptide chain is shown in Figure 7.5.

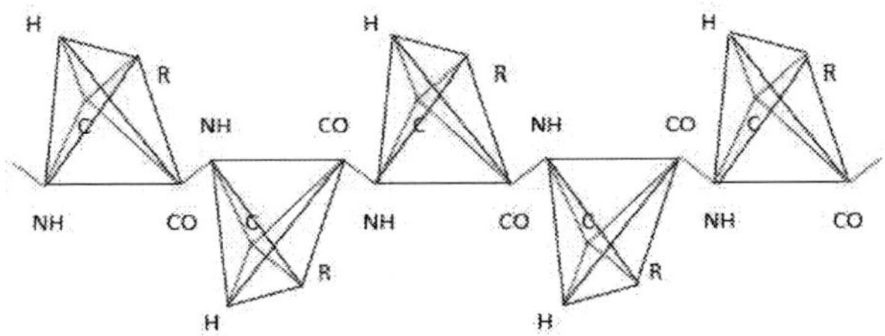

Figure 7.5. Linear polypeptide chain of D-molecules amino acids.

From Figure 7.4 and Figure 7.5 it can be seen that in order for the corresponding vertices of tetrahedrons with the center (i.e., specific atoms or functional groups) could lie on a system of parallel lines (i.e., form a linear polypeptide chain), it is necessary that the amino acid molecules in the chain alternate with their mirror image relative to the edge in the tetrahedron connecting the bond centers of the polypeptide chain (i.e., relative to the segment CO-NH). The chains in Figure 7.4 and Figure 7.5 differ by the mirror reflection of the molecules relative to the perpendicular to the segment *CO-NH* (*L*-and *D*-molecules).

In Figures 7.4 and 7.5, you can see the translational symmetry of the circuits. Moreover, the elementary translational element in this case is a group of two linked amino acid molecules, one of which is a mirror image of another amino acid molecule relative to the segment. From these two linked amino acid molecules, you can create a convex polytope by connecting edges of each vertex of any of the molecules with all other vertices in the group. This is how a simplex polytope of dimension 9 is formed, since the dimension of a simplex polytope is one less than the number of vertices in a simplex (Zhizhin, 2019 a), and the number of vertices in two tetrahedra with a center is 10. Therefore, a polypeptide linear amino acid chain has translation symmetry, in which an elementary translation element is a simplex polytope of dimension 9. In Figures 7.4 and 7.5, as well as in Fischer's projections (Figure 7.1) and in spatial images of molecules (Figure 7.2), in accordance with the definition of functional dimension, functional groups R, CO, NH were used. Analysis of Pauling's works on the structure of protein molecules (Pauling, Corey, 1950; Pauling, et al., 1951; Pauling, Corey, 1953; Corey, Pauling, 1953) shows that Pauling's model is nothing more than an assumption. Consider three amino acid residues linked by a peptide bond (Figure 7.6).

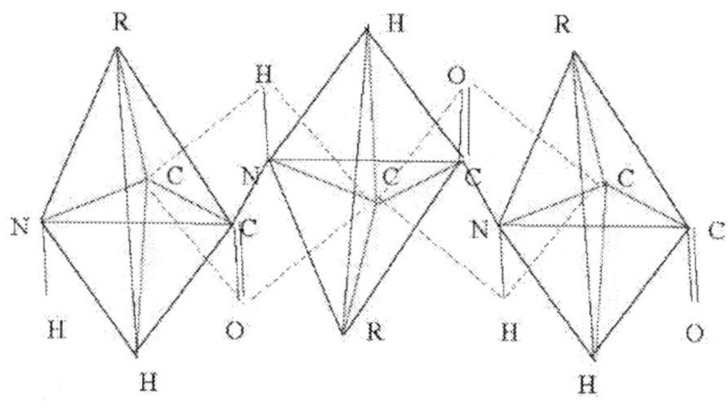

Figure 7.6. Spatial image of three amino acid residues linked by a peptide bond.

Figure 7.6 shows three amino acid residues in the form of tetrahedra, in the center of which there are three carbon atoms, from which four valence bonds proceed to functional groups located at the vertices of the tetrahedra. These valence bonds are marked with red ribs. Three amino acid residues are linked by two peptide bonds located within two contours, indicated by blue dashed lines. According to Pauling's model, the atoms located on the dotted contour and inside the contour are on the same plane. Here we must take into account that the lengths of chemical bonds connecting different pairs of atoms are different. Calculations of bond angles in Pauling's works are not given for any of the bonds. The plane is determined, as is known, by three points. The statement that some other point belongs to a given plane requires a geometric proof. There is no such evidence in Pauling's work with coworkers. Therefore, Pauling's model remains unproven. From the point of view of chemistry and geometry, the tetrahedral coordination of functional groups relative to the central carbon atom in the amino acid molecule is much more important. This fact should be the basis for the analysis of protein structures.

Thus, it is essential (see Figure 7.6) that the α-carbon atoms do not lie in the same plane with the peptide bond atoms. From a comparison of Figures 7.4-7.6 it follows that α-carbon atoms lie either above the plane in which the atoms of the peptide bond are located, or below this plane, and can never lie on this plane. Therefore, Pauling's idea that α-carbon atoms lie in the same plane with functional groups CO, NH is erroneous. Therefore, Pauling's approximation cannot be used both in the construction of linear polypeptide chains and in the construction of more complex conformations of amino acid molecules.

7.3. Turns of Polypeptie Chains

The polypeptide chain can change its direction. This, in particular, is required in the formation of globular proteins. In this case, loops of the polypeptide chain are formed. Since the polypeptide chain, as we have seen, is a chain of polytopes of higher dimension, it is necessary to verify the possibility of such turns in the chain of polytopes of higher dimension, which are a chain of amino acid molecules. Figure 7.7 shows a possible implementation of such a rotation. It is interesting to note that when the polypeptide chain is rotated, the enantiomorphic shape of the amino acid molecules changes.

This follows from the geometric analysis of the rotation presented in Figure 7.7. If the upper part of the polypeptide chain in Figure 7.7 is considered as the initial one, then it is a sequence of D -molecules of the amino acid from left to right. When the chain is rotated, a polypeptide chain is formed, going from right to left (right turn), while the amino acid molecules look like L -molecules. If the lower part of the polypeptide chain in Figure 7.7 is considered the initial one, then the chain is a sequence of L-molecules going from left to right. When turning in this case (left turn), the molecules take the form of D-molecules going in sequence from right to left. The change in the shape of amino acid molecules during the rotation of the polypeptide chain can be considered, in a sense, the justification for the existence of enantiomorphic forms of amino acid molecules.

Changes in the enantiomorphic forms of amino acid molecules during chain rotations have not previously been paid attention.

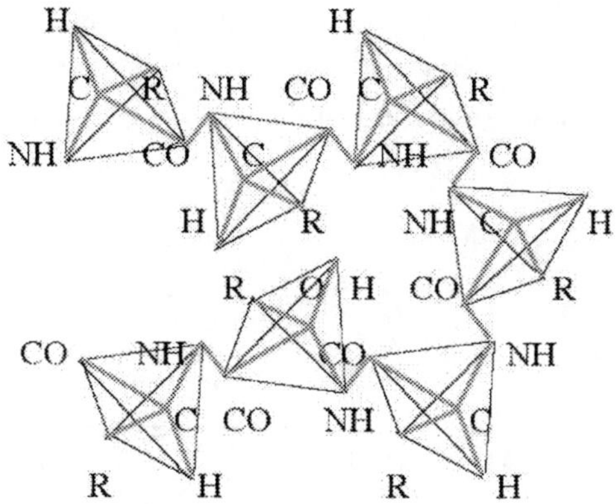

Figure 7.7. The turns of polypeptide chains.

7.4. Spiral Polypeptide Chains

In the previous section, it was proved that the centers of tetrahedrons cannot lie on a flat sheet in which the peptide bond is located. Therefore, Pauling's idea of the connection of such flat leaves at the centers of tetrahedrons to form a helical peptide chain, as Pauling did, cannot be used. There is a relatively simple way of forming a helical polypeptide chain based on the concept of the four-dimensionality of amino acid molecules. This method was proposed in 2016 (Zhizhin, 2016 a). Suppose another amino acid molecule (not in a mirror image) is attached to an amino acid molecule (tetrahedron with a center) by a peptide bond C-N (Figure 7.8) so that in the general case between the valence bond C-C of the first molecule and the peptide bond C-N there is a certain angle less than 180 degrees (since there is no reason to assert that one of these connections is an exact continuation of the other).

Therefore, the second amino acid molecule is rotated by some angle relative to the first amino acid molecule. At the same time, while maintaining the shape of the second amino acid molecule, the second peptide bond, with the help of which the third amino acid molecule can be attached, turns out to be significantly (almost 90 degrees) rotated with respect to the first peptide bond. The third peptide bond is thus almost antiparallel to the first peptide bond. There is a rotation of the attached amino acid molecules relative to some center O (Zhizhin, 2016 a, 2018). If we designate the length of the peptide bond "*a*," the length of the edge of the tetrahedron between the atoms C and N "*b*," then in the projection onto the plane perpendicular to the axis of rotation, we get a polygon (Figure 7.9), which corresponds to the particular with the period of rotation of the molecules equal to 4.

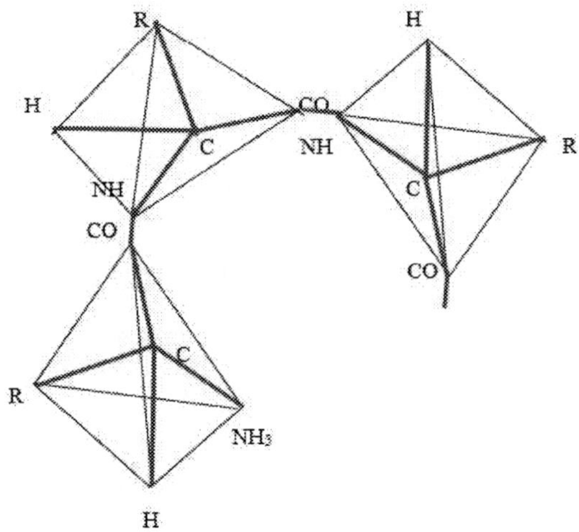

Figure 7.8. Spiral polypeptide chains.

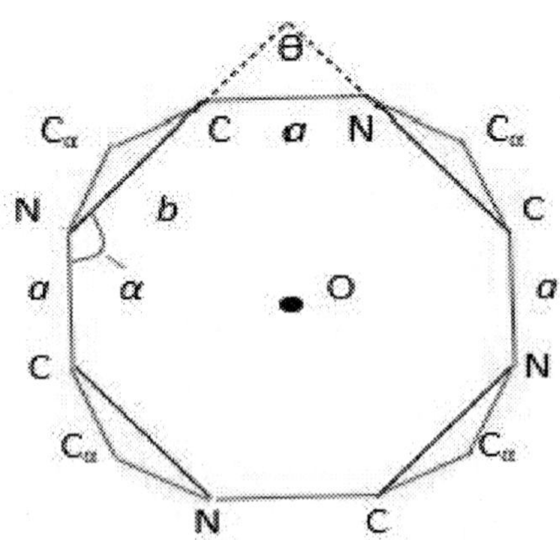

Figure 7.9. Rotation of peptide bonds.

If we continue the nearest sides of the polygon b until they intersect with each other, we find the angle between these extensions. This is the angle $\theta = 180 - 2(180 - \alpha), \alpha$ is the angle between sides a, and b. From here we find the connection between the period n and the angle α

$$n\theta = 360, \alpha = \frac{1}{2}(\frac{360}{n} + 180).$$

If the period $n = 4$ so $\alpha = 135^0$. Pauling had a minimum period of 3.6, but it should be noted that the rotation period of an amino acid molecule can only be an integer. If we now assume that the lines connecting the centers of the tetrahedrons do not lie in the plane of rotation but have a certain angle with respect to this plane, then we get a helicoid of rotation of the amino acid molecule (i.e., α-helix, since α -carbon atoms are located at the centers of the tetrahedrons). It can be shown that if we use the values of the valence bond lengths and the bond angle determined in Pauling's works, then to achieve the distance between the turns of 0.54 nm in the spiral (adopted in Pauling's works), the required value of the angle between the line connecting the centers of the tetrahedra and the plane of rotation, is equal to 1.5 degrees. It should be noted that when constructing a helix, a fundamentally different method of connecting amino acid molecules is used in comparison with a linear peptide chain. This method is more natural since it does not require the use of mirror-opposite amino acid molecules. Between adjacent turns of the spiral of amino acid molecules there is a stabilizing hydrogen bond between the H atoms of the peptide bond and the O atoms of the peptide bond with a double bond.

7.5. Folder Structures of the Amino Acids

One of the important principles of protein structure formation is the formation of as many hydrogen bonds between groups as possible. This is especially pronounced in the so-called β-structures (folder structures) in which elongated chains of amino acid molecules are hydrogen bonded to each other to form a continuous layer. In these structures, the hydrogen bond plays a decisive role, and when calculating the dimensions of the compounds, it must be taken into account on an equal footing with the covalent bond. The repeating element in these β-structures is a polyhedron composed of two amino acid molecules. Figure 7.10 shows one of two variants of such polyhedrons in a fairly arbitrary position.

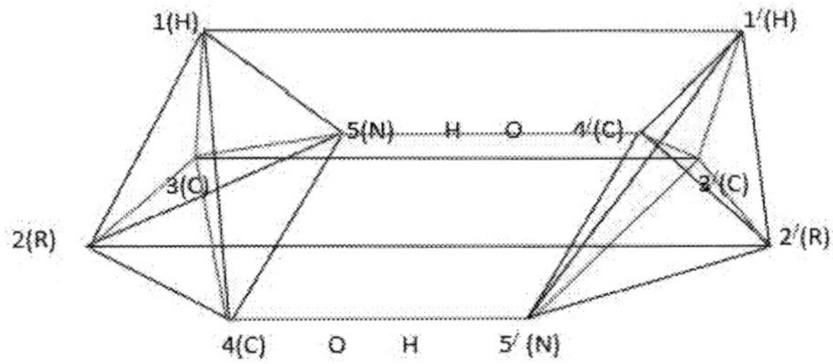

Figure 7.10. Polyhedron composed of two amino acid molecules.

It should be noted that the H and O atoms are not the tops of the figure. Therefore, the polyhedron in Figure 7.10 has 10 vertices ($f_0 = 10$), 24 edges ($f_1 = 24$), 29 two-dimensional faces ($f_2 = 29$), 20 three-dimensional faces ($f_3 = 20$), 7 four -dimensional faces ($f_4 = 7$). Four-dimensional faces are faces 12345, $1'2'3'4'5'$, like tetrahedrons with a center. In addition, tetrahedrons are four-dimensional faces located on four triangular faces of tetrahedron with centers $13451'3'4'5'$, $13251'3'2'5'$, $12341'2'3'4'$, $32543'2'5'4'$. These figures are topologically equivalent to the products of tetrahedrons and a one-dimensional segment (Zhizhin, 2019 a). The difference between these figures and the products of tetrahedron and a one-dimensional segment is only that in this case the one-dimensional segment as a factor in these figures discretely changes its length. But this does not change the number of elements of different dimensions included in the figure. Therefore, as well as the product of a tetrahedron and a one-dimensional segment, these figures include 8 vertices ($f_0 = 8$), 16 edges each ($f_1 = 16$), 14 flat faces ($f_2 = 14$), 6 three-dimensional faces ($f_3 = 6$). Substituting these numbers into the Euler-Poincaré equation (7.1), we obtain

$$8 - 16 + 14 - 6 = 0.$$

This proves that for all the listed figures with vertices 3 and $3'$ the Euler-Poincaré equation is fulfilled for n = 4, i.e., all of them have dimension 4. Another four-dimensional figure is formed by two tetrahedrons 1254 and $1'2'5'4'$ connected by parallel lines. They are also topologically equivalent to the product of a tetrahedron and a one-dimensional segment. Thus, the total number of four-dimensional figures included in Figure 7.10 is 7. Substituting the values obtained for Figure 7.10 into the Euler-Poincaré equation (7.1), we find

$$10-24+29-20+7=2.$$

This proves that in this case the Euler-Poincaré equation is fulfilled for n = 5, i.e., Figure 7.10 has dimension 5.

In a continuous layer of amino acid molecules, the polyhedron in Figure 7.10 alternates with another polyhedron formed by two tetrahedrons with center at other positions (Figure 7.11).

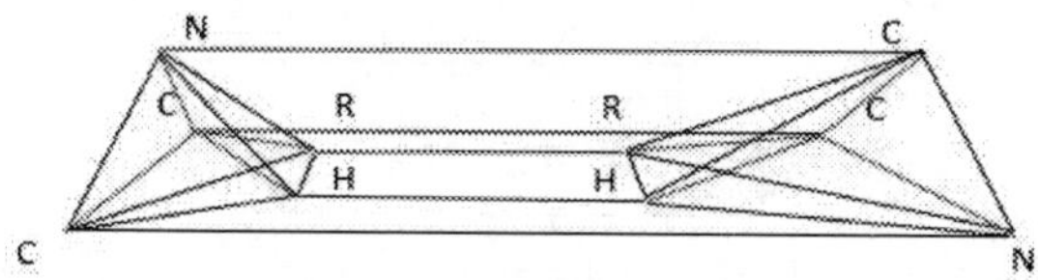

Figure 7.11. Second polyhedron formed by two tetrahedrons with center.

It can be seen that Figure 7.11 is also topologically equivalent to the product of a tetrahedron with center on a one-dimensional segment and therefore has dimension 5. The polyhedrons in Figure 7.10 and Figure 7.11 cannot directly adjoin each other. Between them are other polyhedron formed from two three-dimensional pyramids attached to each other along flat faces. One of these topologically equivalent polyhedrons is shown in Figure 7.12.

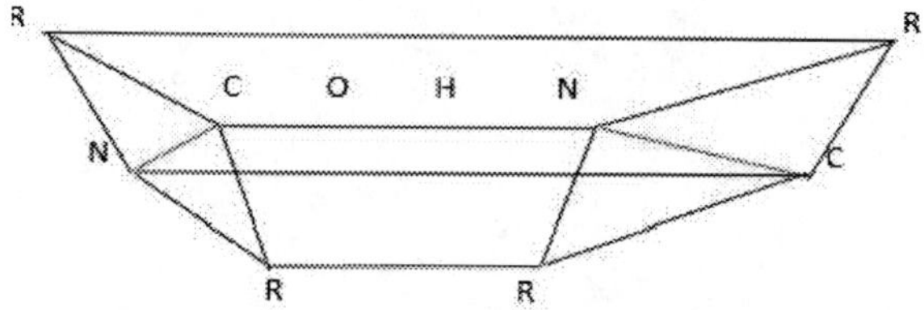

Figure 7.12. Polyhedron of two three-dimension pyramids.

You can see that it has 8 vertices ($f_0 = 8$), 14 edges ($f_1 = 14$), 9 two-dimensional faces ($f_2 = 9$), 3 three-dimensional faces ($f_3 = 3$). Substituting these numbers into the Euler-Poincaré equation (7.1), we find

$$8\text{-}14 + 9\text{-}3 = 0.$$

Therefore, Figure 7.12 satisfies the Euler-Poincaré equations for n = 4, i.e., it has dimension 4. A topologically equivalent Figure 7.13 also has dimension 4.

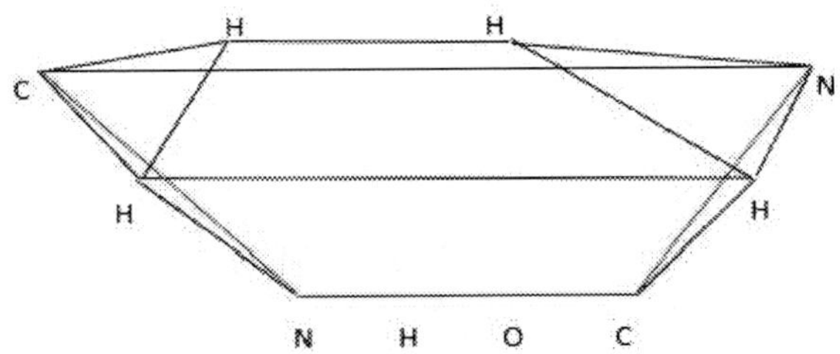

Figure 7.13. Polyhedron of two three-dimension pyramids equivalent Figure 7.12.

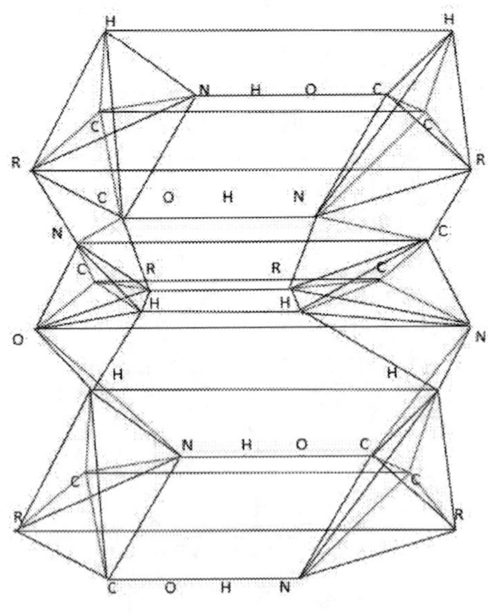

Figure 7.14. Layer composed of periodically repeating polyhedrons 8,9,10, 11.

A part of the common layer composed of periodically repeating polyhedrons 8, 9, 10, 11 is shown in Figure 7.14.

Figure 7.14 shows that the β-structure has translational symmetry. An elementary element of translational symmetry is the sequence of polyhedrons shown in Figures 7.10-7.13. If we denote the polyhedron in Figure 7.10 by symbol 5 (1), the polyhedron in Figure 7.11 by symbol 5 (2), and the polyhedron in Figure 7.12 by symbol 4 (1), the polyhedron in Figure 7.13 by the symbol 4 (2), then the translational element of the β-structure in Figure 7.14 is the polyhedron 5 (1)-4 (1)-5 (2)-4 (2). This polyhedron contains 24 vertices. If we connect each vertex of this polyhedron with all other vertices by edges, we get a single convex simplex polyhedron of dimension 23. Thus, the translational element of the β-structure in Figure 7.14 is a simplex of dimension 23.

In the literature, the β-structures traditionally described are called antiparallel and are depicted as alternating flat two-dimensional stripes (sheets).

However, it is clear that the name anti-parallel in this case does not reflect much of the essence of the structure in Figure 7.14. More precisely, this structure could be called non-parallel. In addition, and this is essential, the β-structures are fundamentally not flat and represent periodically repeating 5 -dimensional and 4-dimensional regions, and each of these regions has two topologically equivalent images.

There are β-structures that include parallel 4-dimensional amino acid molecules with two different orientations. In these structures, alternation of 5-dimensional and 4-dimensional regions (polyhedrons) also occurs. In continuous form, it is presented in Figure 7.15. Figure 7.15 shows that the β-structure has translational symmetry too.

Similar this the β-structure in Figure 7.14, it is also possible to define the sequence of polyhedrons of dimensions 5 and 4, which make up the translational element of the β-structure in Figure 7.15. Given the difference in the form of polyhedrons in Figures 7.14 and 7.15, the sequence of polyhedrons that make up the translational element of the structure in Figure 7.15 can be written as 5 (3)-4 (3)-5 (4)-4 (4). The particular type of polyhedrons of dimensions 5 and 4 is easy to determine from Figure 7.15. The translational element of β-structure in Figure 7.15 also includes 24 vertices and, therefore, is a simplex polyhedron of dimension 23.

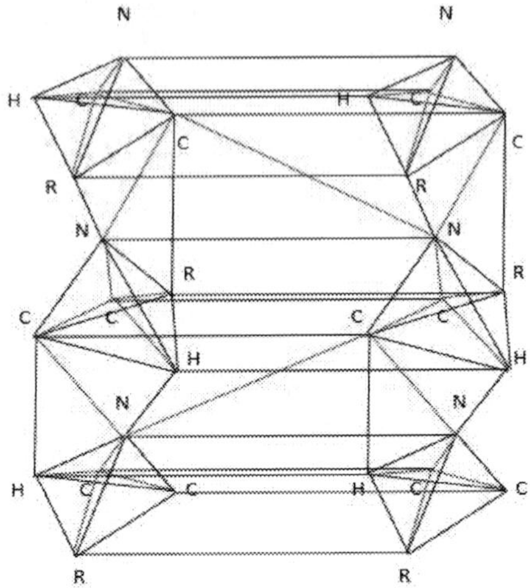

Figure 7.15. Folder structure with parallel molecules amino acid in two different orientations.

Most proteins in a living cell are characterized by a more complex arrangement of amino acid molecules associated with a significant influence of hydrogen bonds and the formation of quaternary structures based on secondary and tertiary structures (i.e., secondary structures with the addition of unstructured fragments). The dimension of such structures (hemoglobin, collagen, etc.) should be yet higher.

Spatial models of the β-structures of protein molecules, forming layers of amino acids, in principle, of unlimited length for both antiparallel and parallel conformation have been constructed. It is shown that the simplified flat Pauling models do not reflect the spatial structure of these layers. Using the recently developed theory of higher-dimensional polytopic prismahedrons, models of the volumetric filling of space with amino acid molecules are constructed. The constructed models for the first time mathematically describe the native structures of globular proteins.

7.6. Native Structure of Globular Proteins with Parallel Arrangement of Amino Acid Residues

A polytopic prismahedron, according to the theory developed in (Zhizhin, 2019 a, b, 2020, 2021 b, c, 2022 b), is a product of two polytopes on top of each other. Moreover, each polytope can have its own dimension. As a result of the product of polytopes, each of them is evenly distributed over the space of their product as a polytope. Therefore, using the theory of polytopic prismahedrons, one can obtain various options for a uniform distribution of polytopes over space in a parallel position. If we take a polytope in an antiparallel position as the initial state, then as a result of the product of polytopes, one can obtain various options for a uniform distribution over space of polytopes in an antiparallel position. In both cases, if the remainder of the amino acid molecule is used as one of the factors, we will obtain uniform spatial distributions of the residues of the amino acid molecule in a strictly parallel or strictly antiparallel position. These distributions will differ from the fairly random distributions discussed earlier in this chapter.

It is known that the residues of amino acid molecules (after the cleavage of water molecules) can be located in a protein in a parallel position (Metzler, 1980; Lehninger, 1982; Koolman, Roehem, 2013). Two such residues, considered together, form a polytopic prismahedron (Figure 7.16).

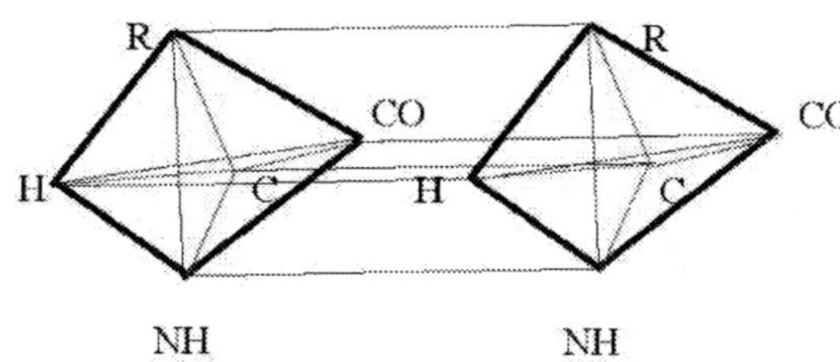

Figure 7.16. The polytopic prismahedron from two residues of amino acid molecules in parallel position.

Its dimension can be determined by equation (7.1). In this prismahedron the number of vertices is 10, the number of edges is 24, the number of two-dimensional faces is 29, the number of three-dimensional faces is 20, the number of four-dimensional faces is 7. In this case $f_0 = 10, f_1 = 24, f_2 = 29, f_3 = 20, f_4 = 7$.

Substituting the obtained values $f_i, i = 0 \div 4$ obtain,

$$10 - 24 + 29 - 20 + 7 = 2.$$

Thus, Euler-Poincaré's equation in this case holds for $n = 5$. This proves that a prismahedron on Figure 7.16 has dimension 5. The polytopic prismahedron in Figure 7.16 is the product of a tetrahedron centered on a one-dimensional segment. The bond between the residues of the amino acid molecule is facilitated by a covalent bond and a hydrogen bond. An example of the resulting structure is shown in Figure 7.17, in which the covalent bond between the residues of the amino acid molecule is indicated by red solid lines, and the hydrogen bond is indicated by red dashed lines.

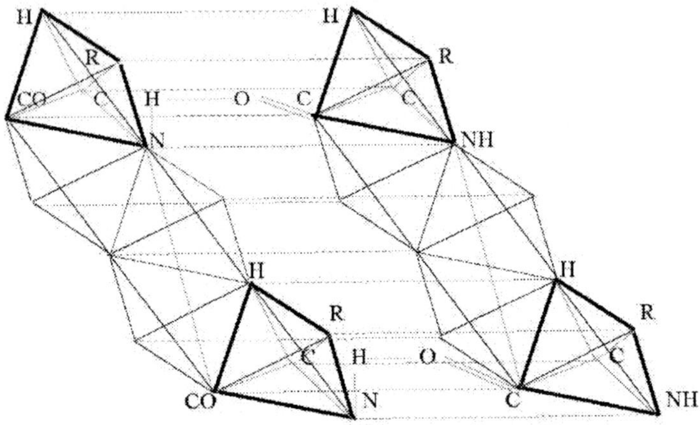

Figure 7.17. The β-structure with parallel residues of the amino acid molecule.

This is how the so-called β-structure in the form of a sheet is formed. However, these sheets are high -dimensional spatial formations. The highest dimension is possessed not only by polytopic prismahedrons with bases in the form of tetrahedra with a center, but also by the figures located between these polytopic prismahedrons. They are polytopic prisms of dimension 4, attached to each other along whole quadrangular faces.

To describe the distribution of residues of amino acid molecules, taking into account their exit from the specified layer, it is necessary to go from the product of a tetrahedron with a center to a one-dimensional segment to the product of a tetrahedron with a center by geometric elements of higher dimension. Consider the product of a tetrahedron centered on a triangle. The process of obtaining the product of a tetrahedron and a triangle was considered earlier (Zhizhin, 2019 a). The product of a tetrahedron with a center and a triangle is obtained in the same way, with the difference that in this case the product result is a polytope of dimension 6. Its image is shown in Figure 7.18.

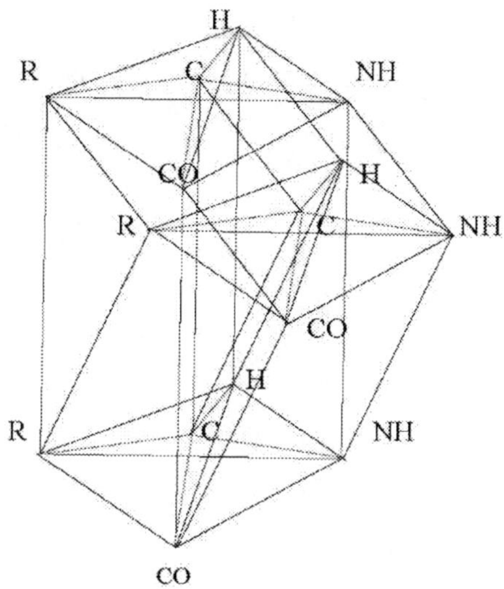

Figure 7.18. The product of the tetrahedron with center by a triangle.

Each of the three pairs of parallel tetrahedrons with a center can lead to the formation of a β-layer of amino acid molecules perpendicular to the forming tetrahedral prism. The space between the residues of amino acid molecules is filled with tetrahedral prisms with tetrahedrons without a central carbon atom. Figure 7.17 shows that the residues of amino acid molecules are distributed in space, moving away from the center of the structure in three directions, forming a globule. (The chemical covalent and hydrogen bonds connecting amino acid residues are not depicted in Figure 7.18 for ease of reference.) Multiplying the tetrahedron centered (the remainder of the amino acid molecule) by the tetrahedron creates a space of dimension 7 (Figure 7.19).

There are 6 pairs of parallel tetrahedrons with a center, i.e., 6 polytopic prisms, each of which forms a β-structure. An even more densely packed globule is thus formed.

Multiplying the tetrahedron with the center (the remainder of the amino acid molecule) by the simplex (dimension 4), we get a polytope of dimension 8 with five residues of the amino acid molecule. This process can be continued, obtaining polytopes of increasing dimensions and with an increasing number of residues of the amino acid molecule. In each case, they will generate periodic β-structures with the formation of ever-increasing globules.

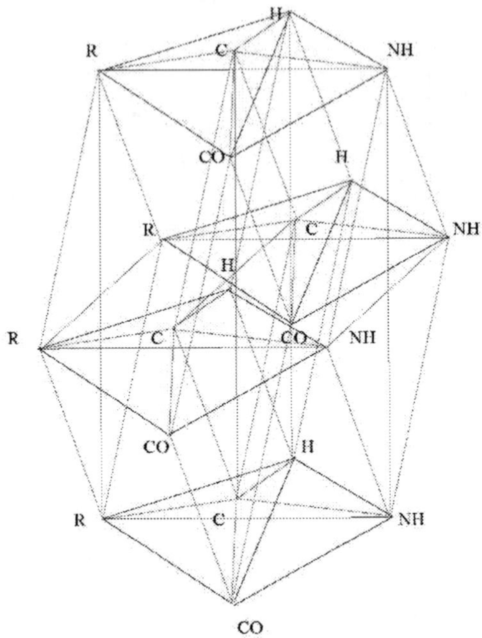

Figure 7.19. The product of the tetrahedron with center by a tetrahedron.

7.7. Native Structure of Globular Proteins with Antiparallel Arangement of Aminoacide Residues

Two tetrahedrons centered in the antiparallel position form a cross-polytope of dimensions 5 (Zhizhin, 2019 a, b). Considering the arrangement of atoms and functional groups at the vertices of the tetrahedron centered on the amino acid molecule, Figure 7.20 shows an image of two residues of the amino acid molecule in the antiparallel position.

It should be noted here that antiparallelism of figures means rotation of one figure relative to another along a helicoidal curve, i.e., in a spiral. Therefore, when depicting the second pair of tetrahedra with centers antiparallel to the first pair, each tetrahedron from this pair also forms a 5-cross-polytope with the corresponding tetrahedron of the first pair of tetrahedra (Figure 7.20).

A closed chain of 5-cross-polytopes with intersection along tetrahedrons with the center is formed. Thus, each tetrahedron with a center participates simultaneously in motion along two helicoidal curves.

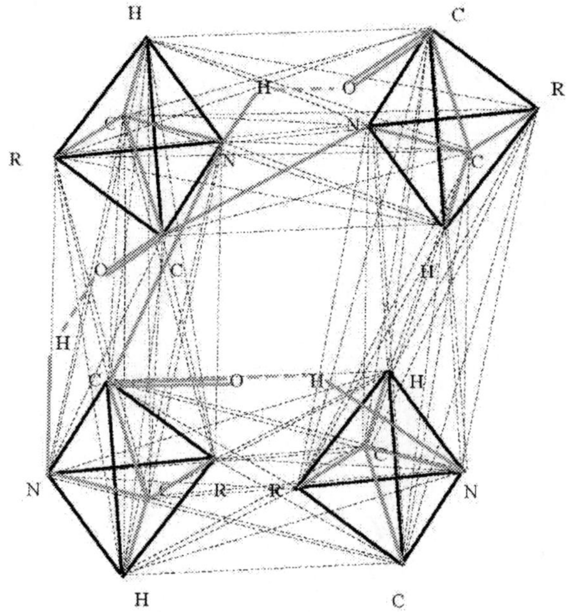

Figure 7.20. Protein structure with antiparallel position of amino acid residues.

If we multiply a construction of four 5-cross-polytopes by a one-dimensional segment, then one more construction will appear, located parallel to the first construction of 5-cross-polytopes. The multiplication process can be continued by increasing the number of tetrahedra with a center (residues of amine acid molecules), the number of 5-cross-polytopes united in chains.

The process of constructing the structure of proteins with antiparallel residues of amino acid molecules naturally leads to the appearance of parallel tetrahedrons in the distribution (see Figure 7.20), since double antiparallelism is parallelism.

However, it is possible to construct a structure of tetrahedrons with center, in which, along with antiparallel tetrahedrons with a center, parallel tetrahedrons with a center from a continuous region (Figure 7.21).

In Figure 7.21, two intersecting 5-cross-polytopes end in two tetrahedrons centered in a parallel position, which form a polytopic prismahedron of dimension 5. In can be continued to form a region of tetrahedrons centered in a parallel position.

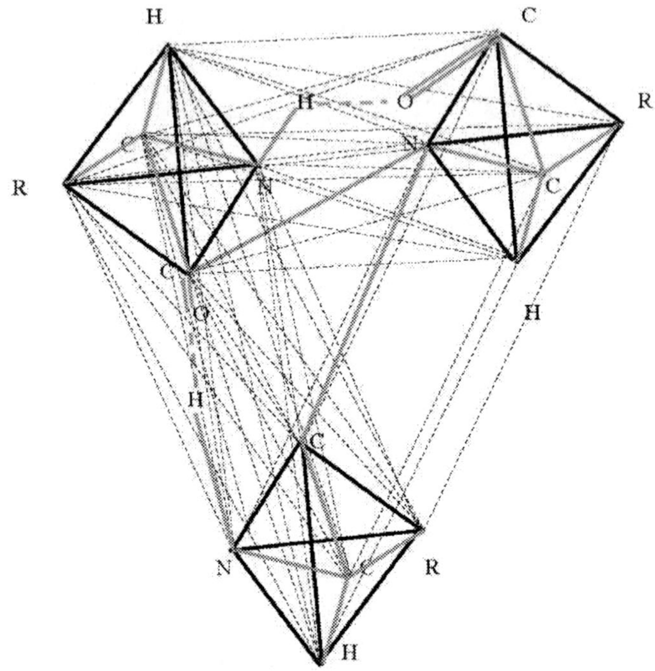

Figure 7.21. Combination of parallel and antiparallel tetrahedrons with a center.

7.8. Native Structure of Globular Proteins with Parallel and Antiparallel Arragement of Amino Acid Residie, and α-Spirals

In order to describe the α-helices existing in the native structure of proteins with the help of tetrahedron with a center, in addition to parallel and antiparallel tetrahedrons with a center, it is necessary to have an intermediate state of a tetrahedron with a center between the two indicated states. It can be obtained by rotating each edge of the tetrahedron centered at 90 degrees, since the antiparallel state to this state is obtained by rotating each edge of the tetrahedron centered at 180 degrees. Each of the polytopes, consisting of two tetrahedra with a center, rotated relative to each other by 90 degrees, have dimension 5. Indeed, let us denote by integers the vertices of any of these polytopes (Figure 7.22).

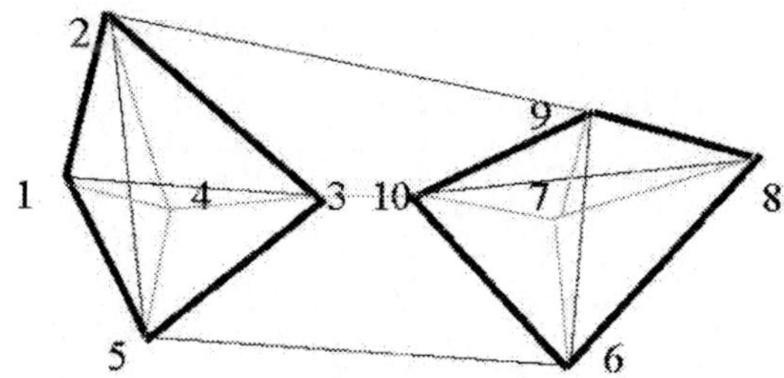

Figure 7.22. Polytope, consisting of two tetrahedra with the center, rotated relative to each other by 90 degrees.

The polytope in Figure 7.22 is not a prismahedron, since the lines connecting the vertices of the two tetrahedra in it are not parallel to each other. This polytope can be called a self-orthogonal polytope. Its dimension can be determined by equation (7.1). In this polytope the number of vertices is 10, the number of edges is 22: 1-2, 1-4, 1-3, 1-5, 2-5, 2-4, 2-3, 3-4, 3-5, 3-10, 4-5, 5-6, 6-10, 6-9, 6-8, 6-7, 7-8, 7-9, 7-10, 8-9, 8-10, 10-9; the number of two-dimensional faces is 22: 1-5-3, 1-2-3, 1-2-5, 1-4-5, 1-2-4, 1-3-5, 2-4-5, 2-3-5, 2-3-4, 3-4-5, 3-5-10-6, 2-3-9-10, 7-9-10, 7-10-8, 7-6-8, 7-6-10, 7-9-6, 6-8-10, 6-8-9, 8-9-10, 6-9-10, 2-5-9-6; the number of three-dimensional faces is 11: 1-2-3-5, 1-4-3-5, 1-2-4-3, 1-2-4-5, 2-4-3-5, 6-8-9-10, 7-9-8-6, 7-8-9-10, 7-8-6-10, 7-9-6-10, 1-2-3-5-6-8-9-10 (polyhedron bounded by the outer surface of the polytope); the number of four-dimensional faces is 3: 1-2-3-4-5, 6-7-8-9-10, 2-4-3-5-10-9-7-6. In this case $f_0 = 10, f_1 = 22, f_2 = 22, f_3 = 11, f_4 = 3$.

Substituting the obtained values $f_i, i = 0 \div 4$

obtain,

$$10 - 22 + 22 - 11 + 3 = 2.$$

Thus, Euler-Poincaré's equation in this case holds for $n = 5$. This proves that a polytope on Figure 7.22 has dimension 5. The four-dimensional polytopes include two tetrahedrons with a center and a polytope 2-4-3-5-10-9-7-6. Let us prove that this polytope has dimension 4. In this polytope the number of vertices is 8, the number of edges is 15: 2-4, 2-5, 2-3, 4-3, 5-3, 3-10, 4-7, 5-6, 2-9, 10-9, 10-7, 10-6, 7-6, 7-9, 9-6; the number of two-dimensional faces is 13: 2-4-5, 2-4-3, 5-4-3, 2-3-5, 10-7-9, 7-9-6, 10-7-6, 10-9-6, 2-3-10-9, 2-4-7-9, 2-5-9-6, 5-4-7-6, 5-3-10-6; the number of three-dimensional faces is 6: 2-3-5-10-9-6, 2-3-4-5, 7-9-10-6, 2-4-5-7-9-6, 5-4-3-10-7-6, 2-3-4-10-7-9. In this case $f_0 = 8, f_1 = 15, f_2 = 13, f_3 = 6$.

Substituting the obtained values $f_i, i = 0 \div 3$

obtain,

8 – 15 + 13-6 = 0.

Thus, Euler-Poincaré's equation in this case holds for $n = 4$. This proves that a polytope 2-4-3-5-10-9-7-6 has dimension 4.

Consequently, all possible polytopes that can be created from two tetrahedrons with a center somehow, polytopic prisms, cross-polytopes and self-orthogonal polytopes, have dimension 5. Each such polytope leads to polytopic prismahedrons, with the smallest dimension 6, with the help of which, in accordance with the theory of polytopic prismahedrons, it is possible to construct a space of the highest dimension.

When this intermediate state and the state antiparallel to it are included in the set of possible states of a tetrahedron with a center, a picture is obtained that describes all the variety of states observed in the native structure of the protein (Figure 7.23).

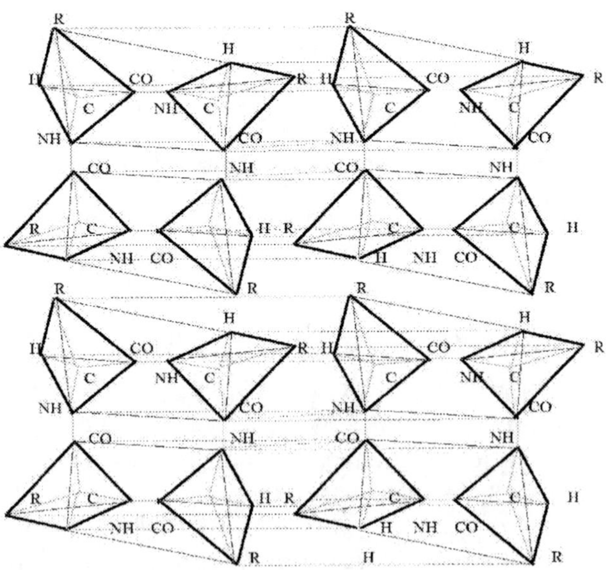

Figure 7.23. Native structure of globular proteins.

If we multiply the picture in Figure 7.23 by geometric elements of different dimensions (segment, triangle, tetrahedron, etc.), we can see how different spatial β-sheets are formed in the native protein structure with parallel and antiparallel arrangement of residues of acid amine molecules, and various α-helices. The presence or absence of certain secondary conformations of protein molecules will be limited by the presence or absence of certain chemical bonds between tetrahedrons centered on a common structure.

7.9. Globular Proteins as Molecular Machines

This class of proteins known as globular proteins that perform complex biological functions. Stabilized by the metal ion contained in them, they control the work of many biological organs, performing the most complex functions, for example, in enzymes and ribosomes. For example, the protein is a globular myoglobin-oxygen-binding protein present in the muscles. In the center of myoglobin globule is hem-group containing Fe-porphyrin (iron atom surrounded by five nitrogen atoms).

7.9.1. Theorem 7.1

The dimension of the Fe-porphyrin before joining the oxygen atom is equal to 5.

7.9.1.1. Proof

Consider the first coordination sphere of the iron atom in the center of the porphyrin (Zhizhin, 2015), since only in the first coordination sphere of atoms are linked by a covalent bond, and in the following focal areas of intermolecular bonds between atoms. Before joining of the oxygen atom, the first coordination sphere of Fe-porphyrin may be represented as a plane projection (Figure 7.24), at the vertices *a, c, d, f* of which the nitrogen atoms of the porphyrin are located, an iron atom is located at the vertex *g*, and the nitrogen atom of the nearest histidine residue is located at the vertex *b*. The deflection of vertex *g* from the center of the rectangle *acdf* corresponds to a certain "dome" character of porphyrin (Steed, Atwwod, 2007; Lehn, 1998). The projection in Figure 7.24 represents some polytope (let's denote *A*-polytope).

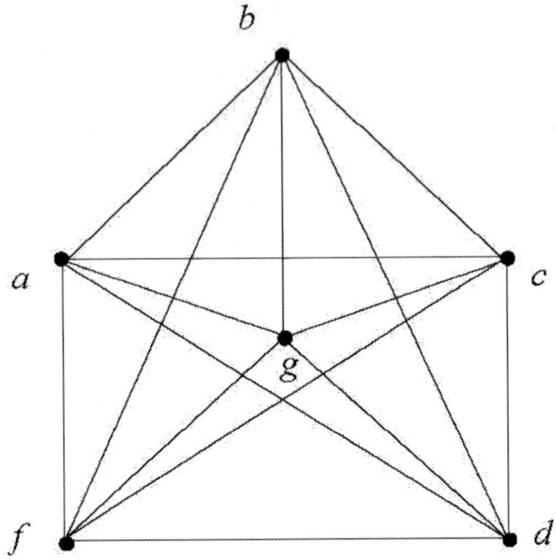

Figure 7.24. The first coordination sphere of Fe-porphyrin before binding oxygen.

The A-polytope has six elements with dimension 0, $f_0(A) = 6$ are vertices *a, c, d, f, g, b*. The number of elements with dimension 1 is $f_1(A) = C_6^2 = 15$. *ab, bc, bd, bf, bg, ac, cd, fd, af, fc, ad, ag, gc, fg*. The number of elements with dimension 2 is $f_2(A) = C_6^3 = 20$. triangles *abf, bfg, bgd, dbc, bga, bgc, agc, dfg, adc, acf, fcd, bgd, fbg, agd, fgc, fbc, abd , afg, gcd, afd*). The number of elements with dimension 3 is $f_3(A) = C_6^4 = 15$. *abgf, bsgd, abfc, abcd, bfcg, abdg, acfg, abdf, acdg, bfdg, abgc, fbcd, fgcd, afgd, afcd*). The number of elements with dimension 4 is $f_4(A) = C_6^4 = 6$. *abcdf, adcdg, abdfg, abcfg, bcdfg, acdfg*. Substituting the received numbers of elements of different dimensions in the equation (2) in Chapter 1 at a value of $n = 5$, we obtain

$$6 - 15 + 20 - 15 + 2 = 2,$$

Thus, the Euler-Poincare equation is satisfied for A-polytope with $n = 5$. This is a simplex of dimension 5. This proves theorem 7.2.

7.9.2. Theorem 7.3

The dimension of Fe-porphyrin after joining of oxygen atom is 6.

7.9.2.1. Proof

The first coordination sphere after joining oxygen atoms is complemented by one vertex *e* (Figure 7.25).

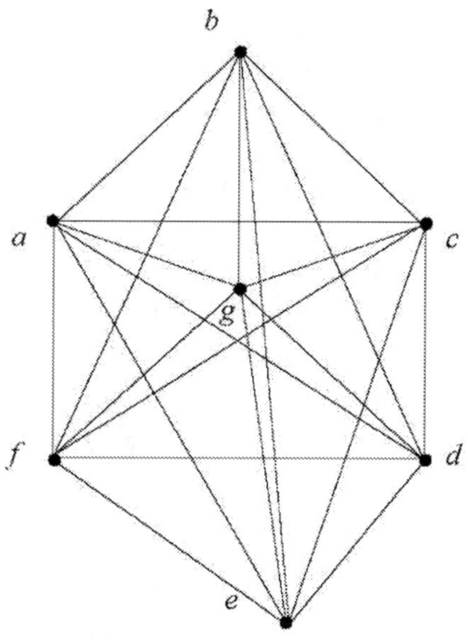

Figure 7.25. The first coordination sphere of Fe-porphyrin after joining of oxygen atom.

"Dome" character of Fe-porphyrin after joining of oxygen atom decreases, but it is not possible to affirm that it disappears completely (Steed, Atwood, 2007). Therefore, the deflection of vertex *g* from the center of rectangle in Figure 7.25 quality is maintained qualitatively. Taking into account the significant difference between the geometry and mass of the groups attached to the iron atom at the top and bottom, it is shown in Figure 7.25 that the vertices *e* and *b* do not lie on the same line. In the polytope in Figure 7.25 (*B*-polytope) the number of elements of zero dimension is increased compared with to the *A*-polytope by one vertex *e*, $f_0(B) = 7$. This leads to the increase in the dimension of the polytope by 1, as the number of edges issuing from each top also increases by 1. In the polytope *B* the number of elements of dimension 1 is $f_1(B) = C_7^2 = 21$ (edges). The number of elements with dimension 2 is $f_2(B) = C_7^3 = 35$ number of three-dimensional figures is $f_3(B) = C_7^4 = 35$ (tetrahedrons). The number of elements with dimension 4 is $f_4(B) = C_7^5 = 21$ of dimension 4). The number of elements with dimension 5 is $f_5(B) = C_7^6 = 7$, (simplexes of dimension 5). Substituting the numbers of the elements of different dimensions in equation (7.1) with $n = 6$, we get

$$7\text{-}21 + 35\text{-}35 + 21\text{-}7 = 0,$$

i.e., the Euler-Poincare equation for *B*-polytope is satisfied when $n = 6$. Therefore, *B*-polytope is a simplex of dimension 6. This proves theorem 7.3.

The dimensions of molecules increase with an increase of its energy again. It is shown that myoglobin is associated coil circuit elements of higher dimension (4) and, moreover, in the center of the coil is a group of atoms even greater dimension.

Conclusion

It is shown that the widespread quasi-plane Pauling models do not reflect and even contradict the spatial structure of polypeptide amino acid molecules, taking into account their highest dimension. This applies to all types of polypeptide chain conformation. In this regard, new models of linear polypeptide chains were construct and it was found that these chains have translational symmetry in the space of higher dimensions. The elementary element of the translational symmetry of these chains is the 9-dimensional polyhedron, which consists of two mirror-opposite amino acid molecules. Helical conformations of polypeptide chains are also composed of higher-dimensional elements. The rotation of monomeric amino acid units in a helical conformation is determined only by the alternation of active centers in an asymmetric amino acid molecule. The study of the amino acid β-structures showed that they also have translational symmetry in the space of the highest dimension. It was found that the β-structures can have various forms, including not containing amino acid molecules in either parallel or antiparallel positions. Moreover, β-structures have elements of translation symmetry with dimension 23, consisting of two topologically equivalent polyhedrons of dimension 5 and two topologically equivalent polyhedrons of dimension 4. The resulting polyhedrons in high dimension, as translation elements of the construction of spaces protein molecules, should be further compared with the general theory of normal partitioning of *n*-dimensional spaces and corresponding high-dimensional parallelohedrons. The turn of the polypeptide chain, which is often found in globular proteins, has been studied in the higher dimensional space. It was found that the rotation of the polypeptide chain is accompanied by a change in the enantiomorphic form of the protein molecules.

The deposition of protein molecules in the brain with the formation of β-amyloid is the cause of many diseases of the nervous system. Until now, the possible structure of the β-amyloid complex and methods of treating these diseases have been little studied. In this work, with the aim of constructing mathematical models of the volumetric filling of space with protein molecules, this possibility is investigated without using simplified flat Pauling models. Analyzing the process of attachment of amino acid molecules under these conditions, it was found that the formed layers of molecules have the highest dimension both in antiparallel and parallel conformation. Based on the recently developed theory of polytopic prismahedrons (Zhizhin, 2019 a, 2021 b, c), a mathematical model of the volumetric filling of space with amino acid molecules was built, which for the first time makes it possible to describe the structure of plaques of protein molecules in the brain. This model can serve as a mathematical model of the native conformation of a globular protein in the general case. Obviously, the specific type of the native structure of the globular protein depends on the type of protein molecules and the conditions for the formation of the globule.

Chapter 8

Geometry of the Structure of Nucleic Acids in the Space of the Highest Dimension

Abstract

The geometrical cause of the formation of different form by molecules nucleic acids (right and left spirals with different number of α-D-ribose and 2-deoxy-α-D-ribose molecules in the period, including closed chains) has been determined. Substituting the known effective values of the lengths of chemical bonds (carbon-carbon, oxygen-oxygen, phosphorus-oxygen) into the structure of polytopes, the values of the characteristic geometric parameters of molecules nucleic acids were calculated: their effective diameter and period. It turned out that the calculated values of these parameters are in good agreement with their values, determined earlier experimentally. It is shown that the set of single-stranded nucleic acids (both DNA and RNA) is broken into two sets of chiral forms. Each form in one set contains a chiral form in another set. Moreover, in each set there are possible rotation of the spirals both in the right and in the left direction.

Keywords: atom, molecule, covalent bond, polymer, simplex

Introduction

In the middle of the last century, there was an idea that nucleic acids (DNA and RNA) ensure the transfer of hereditary information from one generation of living organisms to another (Caspersson, et al., 1941; Brachet, 1942). Moreover, DNA is the main carrier of genetic information, and RNA is an intermediary that receives this information from DNA and implements it in the form of protein biosynthesis. Deoxyribonucleic acid (DNA), as a chemical substance, it was isolated by Johann Friedrich Micher in 1869 from the remains of cells contained in the pus. He singled out a substance that includes nitrogen and phosphorus. When Misher determined that this substance has acid properties, the substance was called nucleic acid (Dahm,

2005). Gradually it was proved that it was DNA, and not proteins, as previously thought, and which is the carrier of genetic information. One of the first decisive proofs was the experiments of Oswald Avery, Colin MacLeod and McLean McCarthy (1944) on the transformation of bacteria.

The structure of the double helix DNA was proposed by Francis Crick and James Watson in 1953 on the base of the X-ray structural data obtained by Maurice Wilkins and Rosalind Franklin and the "Chargaff rules" according to which in each DNA molecule the strict relationships connecting the quantity of nitrogenous bases are different (Watson, Crick, 1953 a, b). For outstanding contributions to this discovery, Francis Crick, James Watson and Maurice Wilkins were awarded the 1962 Nobel Prize in Physiology or Medicine. Deoxyribonucleic acid (DNA) is a biopolymer, the monomer of which is the nucleotide (Albert, et al., 2002; Butler, 2005). Each nucleotide consists of a phosphoric acid residue attached to sugar deoxyribose, to which one of the four nitrogen bases is attached also. The bases that make up the nucleotides are divided into two groups: purines (adenine [A] and guanine [G]) and pyrimidines (cytosine [C] and thymine [T]) are formed by combined five-and six-membered heterocycles.

They managed to show that the DNA isolated from the pneumococci corresponds to the so-called transformation (the acquisition of pathogenic properties by a harmless culture as result of the addition of dead pathogenic bacteria to it). The experiment of American scientists Alfred Hershey and Martha Chase (Hershey-Chase experiment, 1952) with radioactively labeled proteins and bacteriophage DNA showed that only the phage nucleic acid is transmitted to the infected cell, and the new generation of phage contains the same proteins and nucleic acid, as the initial phage (Hershey, Chase, 1952). Deciphering the structure of DNA (1953) has become one of the turning points in the history of biology.

In 1986, Frank-Kamenetskiy in Moscow showed how a double-stranded DNA folds into a so-called H-shape, composed not of two but three strands of DNA (Frank-Kamenetskiy, 1986, 1988). Deoxyribonucleic acid (DNA) is a biopolymer, the monomer of which is the nucleotide (Albert et al., 2002; Butler, 2005).

Nucleotides are long polynucleotide chains covalently linked. These chains in the overwhelming majority of cases (except for some viruses possessing single-stranded DNA genomes) are combined pairwise by means of hydrogen bonds into a secondary structure, called the double helix (Watson, Crick, 1953 a, b; Berg, Tymoczko, Stryer, 2002). Each base at one of the chains is connected to one definite base on the second chain. This specific binding is called complementary. Purines are complementary to pyrimidines (that is, they are capable of forming hydrogen bonds with them): adenine forms bonds only with thymine, and cytosine-with guanine. In a double helix, chains are also linked by hydrophobic interactions and stacking, which do not depend on the DNA base sequence (Ponnuswamy, Gromiha, 1994). Complementarity of the double helix means that the information contained in one chain is also contained in another chain. Different base pairs form a different number of hydrogen bonds. In the future, the existence of nucleic acids differing in the length of the period and shape with rotation of the spiral both to the right and to the left was experimentally established (Ha, et al., 2005; Cantor, Schimmel, 1980; Frank-Kamenetskiy, 2010).

Watson and Crick postulated the spiral form of the DNA molecule, but they did not discuss the reasons for the formation of such a DNA molecule. Until now there have been no works explaining the existence of a spiral in the DNA molecule.

In the book (Zhizhin, 2018), the molecules of practically all the elements of the periodic system were studied in detail and it was shown that many of them, including magnesium and calcium, form compounds of higher dimension (see also Zhizhin, 2015 a, b; Zhizhin, Diudea, 2016; Zhizhin, Khalaj, Diudea, 2016). In the works (Zhizhin, 2016 a, b), the structures of biomolecules, which also form compounds of higher dimension, are also studied. When analyzing the metric space of nucleic acid structures, one should use the representations of their constituents in the form of polytopes of higher dimension. This primarily applies to the phosphoric acid residues and the *D*-ribose molecules, since nitrogenous bases have a dimensionality of 2 and do not have a significant effect on the geometric structure of nucleic acids.

The determination of the geometric forms of chemical compounds according to the Euler -Poincaré (Poincare, 1895) relation translates classical stereochemistry in three-dimensional space into sulfur chemistry in a space of higher dimension. Here new discoveries are possible in the laws of chemical compounds. Consideration of the geometry of molecules using spaces of higher dimension makes it possible to explain the facts observed in reality, which previously had no clear explanations. In the book (Zhizhin, 2018), for example, showed that the rotation of the plane of light polarization in tartaric acid is due to the difference in the shapes of the molecules of *D*-tartaric acid and *L*-tartaric acid, which is clearly visible in their images in five-dimensional space, since their dimension is five. Pasteur (Pasteur, 1960) also tried to explain the rotation of the polarization plane of light in liquid tartaric acid by the difference in the structures of the crystals of *D*-tartaric acid and *L*-tartaric acid. This book received an explanation for the fact that there is no branching in the chain of glucose molecules, if they are bound in β-glycoside linkage. If these molecules are bound in α-glucoside linkage, branching of the chain is possible. It is shown that, because of the higher dimensionality of glucose molecules in the case of a β-glycoside linkage, the centers of possible branching of the chain are blocked by glucose molecules. In this time, with α-glycoside linkage the possible branching points of the chain remain free.

The concept of the geometric correspondence of the structure of substances in the formation of a chemical compound is essential for the form of the compound. However, the complementarity of DNA spirals when connecting spirals cannot be the reason for the formation of these spirals. The geometric shape of the constituent parts of the spiral itself must be important here. They are phosphoric acid residues and *D*-ribose molecules. This paper is devoted to the investigation of this problem. In this case it is necessary to take into account the dimension of these parts, which, as shown in (Zhizhin, 2018), is higher than three. It should be noted that there is no contradiction in the representation of the structure of DNA as a spiral in a three-dimensional space with the highest dimension of its components (Zhizhin, 2019; 2021). The point is that according to the geometry of Riemann (Riemann, 1854), a space of higher dimension has a boundary. Therefore, a space of a smaller dimension can be located outside this space.

The higher dimensionality of molecules can be considered as an example of Big Data. The concept of Big Data has become widespread in computer science in recent times (Kosrapohr, 2003). The representation of molecules in the form of polytopes of higher dimension means the handling of Big Data as geometric images of a set of points in a space of higher dimension. A striking example of the use of the concept of Big Data in medicine is a work of J.A. Rodger (Rodger, 2015) in which, using the intellectual analysis of Big Data, a program has been developed to improve the prognosis of survival rates for traumatic brain injury and other injuries on several ships. The totality of injury data for each vessel can be considered as Big Data in view of the many parameters characterizing trauma.

8.1. The Dimension of Phosphoric Acid and Its Residue

The phosphoric acid H_3PO_4
Karapetians, Drakin, 1994) (Figure 8.1).

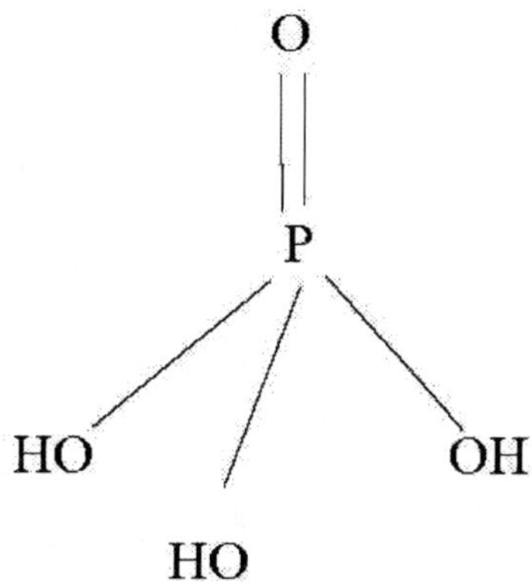

Figure 8.1. Special structure of phosphoric acid.

On the Figure 8.2 is presented Fisher`s formula of phosphoric acid.

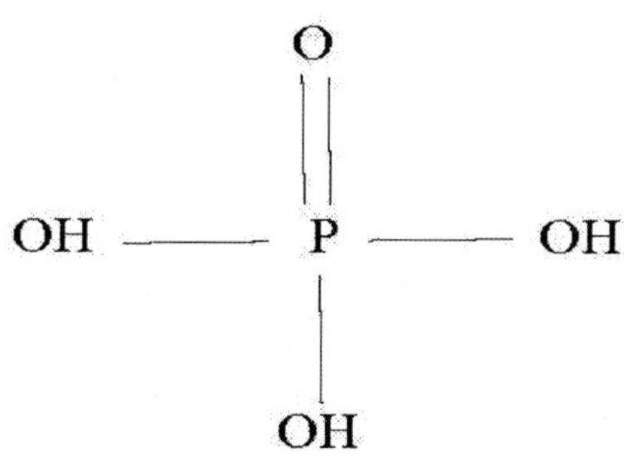

Figure 8.2. Fisher`s formula of phosphoric acid.

It is obvious that, used in the DNA molecule, the phosphoric acid residue obtained by detaching one hydrogen atom from it has the building also. In this case, instead of one of the hydroxyl groups attached to the phosphorus atom, there will be a negatively charged oxygen atom. To determine the functional topological dimension of phosphoric acid and its residue, it is necessary to connect the vertices, atoms and functional groups entering into these compounds (Zhizhin, 2016 b, 2018). Then the structure of phosphoric acid and its residue is represented in the form of a polytope in Figure 8.3.

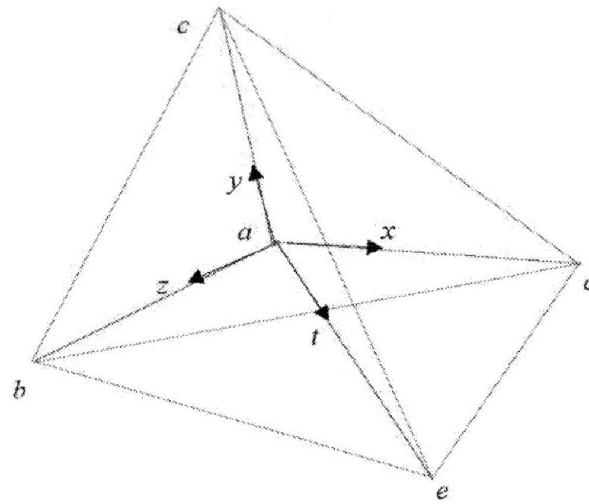

Figure 8.3. The polytope of the phosphoric acid and its residue.

In Figure 8.3 in vertices *a* dispose atom phosphoric; in vertices *c* dispose atom of oxygen with double bond; in vertices *b, d, e* dispose atoms of oxygen with free bond.

Calculation of the functional topological dimension by the Euler-Poincaré equation (Poincaré, 1895) does not require specifying the exact lengths of polytope edges, and for a given number of the vertices does not depend on the groups of atoms in these vertices (Zhizhin, 2016 b). In Figure 8.3, the edges corresponding to the covalent chemical bonds are marked in red, and the edges that carry only the geometric function, delineating the closed geometric figure, are marked in blue.

It follows from Figure 8.3 that the polytope of phosphoric acid has five vertices (*a, b, c, d, e*), ten edges (*ac, ab, ad, ae, cd, de, be, ce, cb, bd*), ten plane triangular faces (*cbe, cda, cea, dab, dbe, dce, bae, dae, bce, dcb*), and five three-dimensional faces in the form of tetrahedron (*cbae, caed, cbad, abde, cebd*). These values of elements of different dimensions we must substitute into the Euler-Poincare equation

$$\sum_{i=0}^{n-1} f_i (-1)^i = 1 - (-1)^i, \qquad (8.1)$$

where f_i i in polytope with dimension n. In this case there are $f_0 = 5, f_1 = 10, f_2 = 10, f_3 = 5$

Substituting these values into equation (8.1), can obtain

5-10 + 10-5 = 0.

Thus, equation (8.1) is satisfied for $n = 4$. This proves that the polytope of phosphoric acid and its residue has a dimension of four. It is worth emphasizing that the polytope of phosphoric acid and its residue is topologically equivalent to a convex polytope of the simplex type with dimension four, since these polytopes completely coincide with the values of the numbers of elements of different dimensions (Zhizhin, 2014 b). The origin of the four-dimensional space can be placed in one of the vertices with sending coordinates along the edges connecting this vertex to other vertices (see Figure 8.3).

8.2. The Dimension of the Molecules Deribose and Deoxyribose

The *D*-ribose is molecule of carbohydrate with five atoms carbon. There are two enantiomer forms of this molecule. Its images in the form of Fisher's formula (Metzler, 1980) are shown in Figure 8.4 A, B.

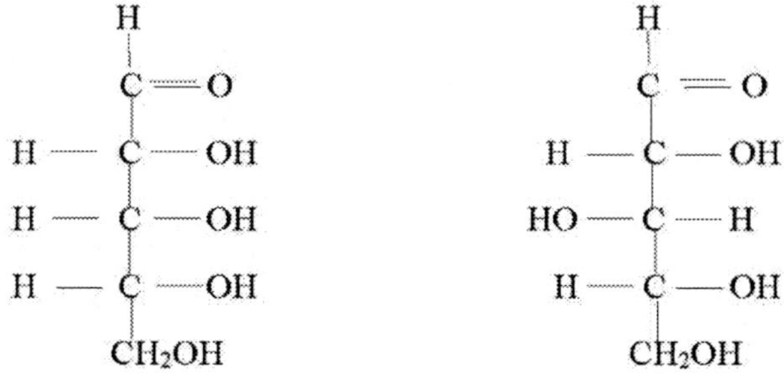

A) B)

Figure 8.4. A, B. Two enantiomer forms of the molecule D-ribose (formulas of Fisher's).

In the DNA and RNA molecules, *D*-ribose as a component enters with a closed carbon chain (in form furanose). Moreover, in DNA molecules, D-ribose takes the form of deoxyribose, in which the hydroxyl group at the second carbon atom in the closed chain is replaced by a hydrogen atom (2-deoxy-D-ribose). The closed chain of the molecule in Figure 8.4 is formed due to the rupture of the double bond between oxygen and carbon (considering it the first in the chain), the liberation of the water molecule, and the addition of this oxygen atom to the penultimate (fourth) carbon atom in the carbon chain. The Fischer formula or formula Haworth and their modifications (Metzler, 1980; Lehninger, 1982) may not reflect the spatial structure of the *D*-ribose molecule. For this target the constructs described in the form of convex polytopes with boundary elements which form boundary complex (Grunbaum, 1967). Figure 8.5 shows the result of the closure of a chain of four carbon atoms through an oxygen atom correspondent Figure 8.4 A. Chemical bonds are marked as before with red color.

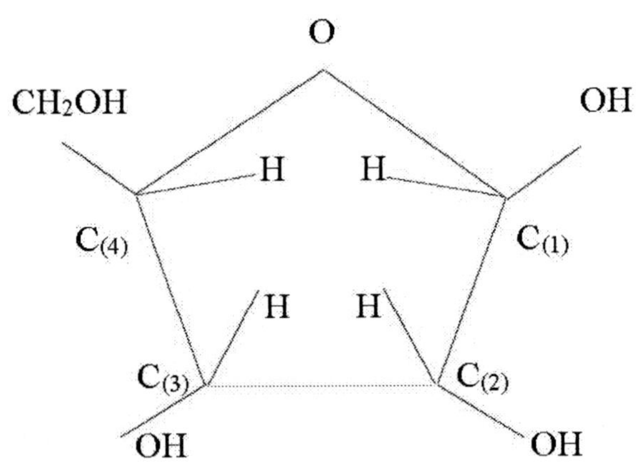

Figure 8.5. The molecule α-D-ribose with closed a chain of carbon atoms (the form A).

It is easy to see that, in addition to the form of the α-*D*-ribose molecule in Figure 8.5 (a form of *A*), there is another enantiomer form of this molecule. It connected by a symmetry transformation with respect to a straight line passing through an oxygen atom separating the projection of the form of *A* in half (we shall assume that it is a *B* shape). Form *B* of the α-*D*-ribose molecule is shown in Figure 8.6.

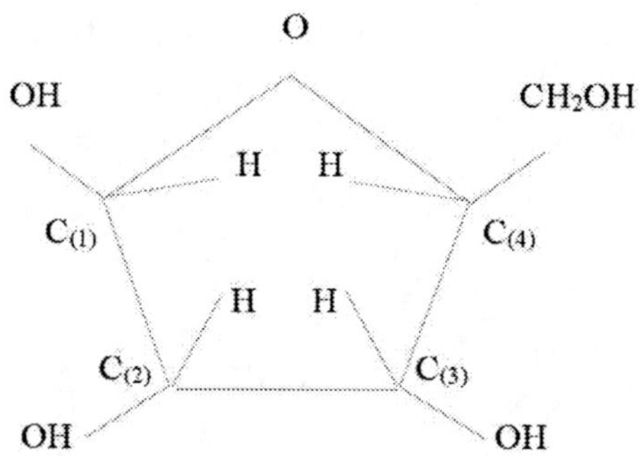

Figure 8.6. The molecule α-D-ribose with closed a chain of carbon atoms (the form B).

Figure 8.7 shows the result form of the molecule β-D-ribose with closed a chain of carbon atoms (the form A).

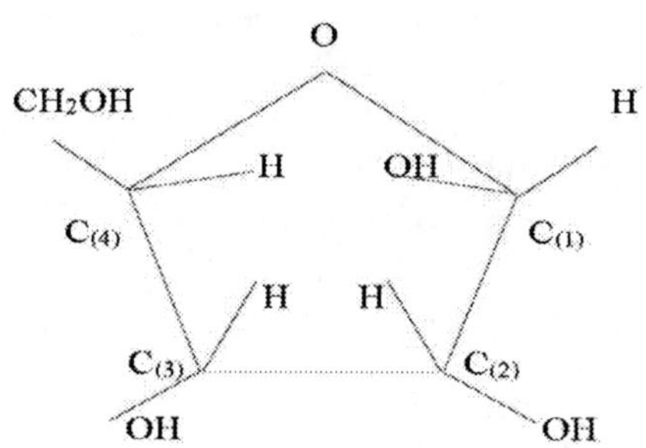

Figure 8.7. The molecule β-D-ribose (A).

Figure 8.8 shows the result form of the molecule 2-deoxy-α-D-ribose with closed a chain of carbon atoms (the form A).

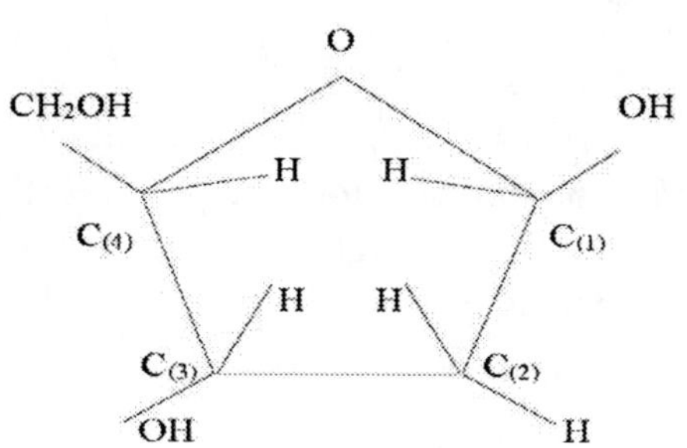

Figure 8.8. The molecule 2-deoxy-α-D-ribose (A).

Figure 8.9 shows the result form of the molecule 2-deoxy-β-D-ribose with closed a chain of carbon atoms (the form A).

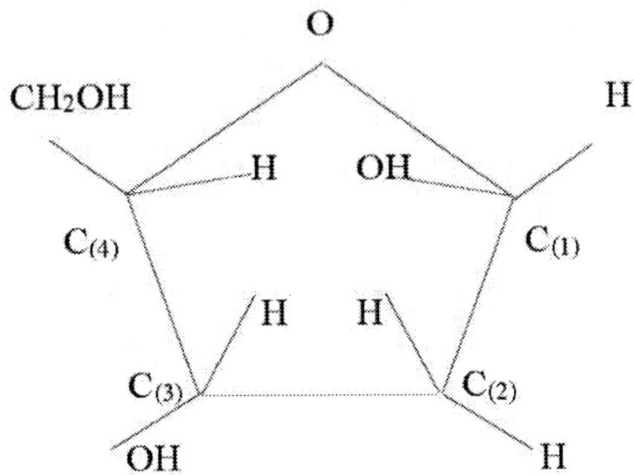

Figure 8.9. The molecule 2-deoxy-β-D-ribose (A).

When depicting molecules in Figures 8.5-8.9, we transfer the hydroxyl groups associated with carbon atoms in the chain to the outer contour of the molecules, given that it is the hydroxyl groups that ensure the implementation of chemical reactions of sugar molecules that really exist, in particular, When depicting molecules in Figures 8.5-8.9, we transfer the hydroxyl groups associated with carbon atoms in the chain to the outer contour of the molecules, given that it is the hydroxyl groups that ensure the implementation of chemical reactions of sugar molecules that really exist, in particular, in the formation of nucleic acids. Obviously, for each image in Figures 8.7-8.9 can be obtained by transforming the symmetry of the molecule in the form B, just as it was obtained for Figure 8.6.

In order to give the molecules on Figures 8.6-8.9 a kind of polytopes, it is necessary to connect on this Figures each vertex (every atom or functional group, with all other vertices) to the ribs, then can get polytopes of the simplex type, the dimension of which must be determined.

For the α-D-ribose molecule (Figure 8.5), this was done in Chapter 6 (Figure 6.10). In this chapter, we will represent a molecule 2-deoxy-α-D-ribose (A) in the form of a polytope (Figure 8.10).

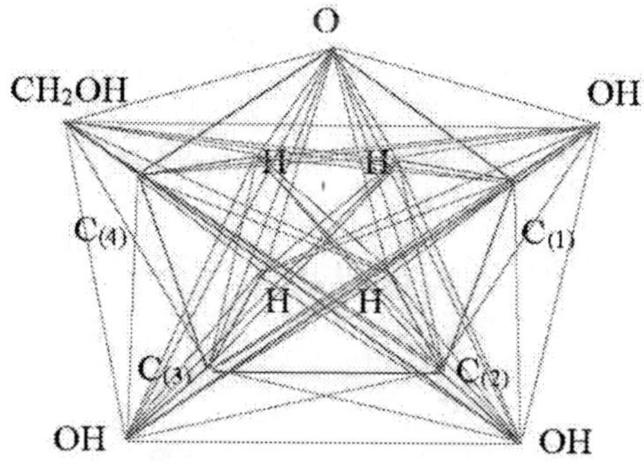

Figure 8.10. The molecule 2-deoxy-α-D-ribose (A) in form of the polytope.

The polytope at Figure 8.10 has 13 vertices in which there are one oxygen atom, four hydrogen atoms, three hydroxyl groups, four carbon atoms, compound CH_2OH. It is known that the dimension of a simplex is one less than the number of its vertices (Zhizhin, 2019). Therefore, the dimension of the polytope in Figure 8.10, if the Euler-Poincaré equation (8.1) holds for it, equals $n = f_0 - 1 = 12$

polytope, and all its faces therefore are simplexes. Since all the vertices of a simplex are related to each other, the number of edges is determined by the number of combinations of the number of vertices of two. In this case, the number of edges of the polytope in Figure 8.10 is $f_1 = C_{n+1}^2 = 78$. The number of two-dimensional faces (triangles) of this polytope is $f_2 = C_{n+1}^3 = 286$.

The number of three-dimensional faces (tetrahedrons) of this polytope is $f_3 = C_{n+1}^4 = 715$.

The number of four-dimensional faces (4-simplexes) of this polytope is $f_4 = C_{n+1}^5 = 1287$.

The number of five-dimensional faces (5-simplexes) of this polytope is $f_5 = C_{n+1}^6 = 1716$.

The number of six-dimensional faces (6-simplexes) of this polytope is $f_6 = C_{n+1}^7 = 1716$.

The number of seven-dimensional faces (7-simplexes) of this polytope is $f_7 = C_{n+1}^8 = 1287$. The number of eight-dimensional faces (8-simplexes) of this polytope is $f_8 = C_{n+1}^{9} = 715$. The number of nine –dimensional faces (9-simplexes) of this polytope is $f_9 = C_{n+1}^{10} = 286$.

The number of ten-dimensional faces (10-simplexes) of this polytope is $f_{10} = C_{n+1}^{11} = 78$.

The number of eleven-dimensional faces (11-simplexes) of this polytope is $f_{11} = C_{n+1}^{12} = 13$. Substituting the values $f_i, (0 \le i \le 11)$ in the Euler-Poincaré equation (8.1), can see that it holds for n = 12 for polytope on Figure 8.10

$$13-78 + 286-715 + 1287-1716 + 1716-1287 + 715-286 + 78-13 = 0.$$

This finally proves that the polytope 2-deoxy-*α*-*D*-ribose (A) has dimension 12 and a simplex type. Obviously, that all polytopes α-*D*-ribose (A), *β*-*D*-ribose (A), *α*-*D*-ribose (B), *β*-*D*-ribose (B), 2-deoxy-α-D-ribose (A) , 2-deoxy-*α*-*D*-ribose (B), 2-deoxy-*β*-*D*-ribose (A), 2-deoxy-*β*-*D*-ribose (B) has the same dimension and type, as all have equal the numbers $f_i, (0 \le i \le 11)$.

8.3. The Structure of the *α*-*D*-Ribose and 2-Deoxy *α*-*D*-Ribose Nucleic Acids

The nucleic acid is a chain of *D*-ribose (α-*D*-ribose or 2-deoxy-*D*-ribose) molecules between which are the phosphoric acid residues that connect these molecules. Let us construct this circuit, taking into account the concept of repulsion of electron pairs (Gillespie, 1972; Gillespie & Hargittai, 1991). According to this concept, tetrahedral coordination of atoms and functional groups in these compounds is carried out around the phosphorus atom in the phosphoric acid residue, and in the carbon atoms in the *D*-ribose molecule also. It is this concept that leads to the higher dimensionality of the remainder of phosphoric acid and the molecule of *D*-ribose. When a phosphoric acid residue is added to the *D*-ribose molecule from the hydroxyl group of the phosphoric acid residue and the hydroxyl group of the *D*-ribose molecule, a water molecule that leaves the compounds forms and an oxygen atom that binds the phosphoric acid residue and the *D*-ribose molecule. The concept of repulsion of electron pairs also operates here, according to which two valence electrons of the oxygen atom maximally repel each other, forming electron pairs with electrons of the phosphorus atom and the carbon atom. Therefore, taking into account that tetrahedral coordination is carried out around the atoms of phosphorus and carbon already, it can assume that the valence bonds of the oxygen atom are located linearly with respect to the oxygen atom. When the second *D*-ribose molecule is added to the phosphoric acid residue, the second hydroxyl group of the phosphoric acid residue with the compound CH_2OH of this *D*-ribose molecule also form molecules of water that leaves the compound and an oxygen atom connecting the phosphorus atom and the functional group CH_2 of the *D*-ribose molecule. By virtue of the concept of repulsion of electron pairs, the linear arrangement of the valence bonds of the oxygen atom can be considered here also. In addition, as the first step within the framework of the notions of the functional dimension of molecules (Zhizhin, 2016 b), can shall assume the linear arrangement of the valence bonds also in the neighborhood of the functional group CH_2 of the *D*-ribose molecule. One will take into account this when constructing a chain of *D*-ribose molecules and phosphoric acid residues. To construct the sequence of *D*-ribose molecules and the phosphoric acid residue in the chain can will use a simplified image of the *D*-ribose molecule of dimension 12. Let us leave only the outer contour and valence bonds necessary for interaction with the environment from the *D*-ribose molecule. At the same time, one will not depict the hydrogen atoms inside the contour of carbon atoms. We approximately to consider that the lengths of chemical bonds carbon-carbon, carbon-oxygen close to each other. The images of the α-*D*-ribose and 2-deoxy-α-D-ribose molecules of the enantiomer form *A* under these conditions takes the same form (Figure 8.11 and Figure 8.12).

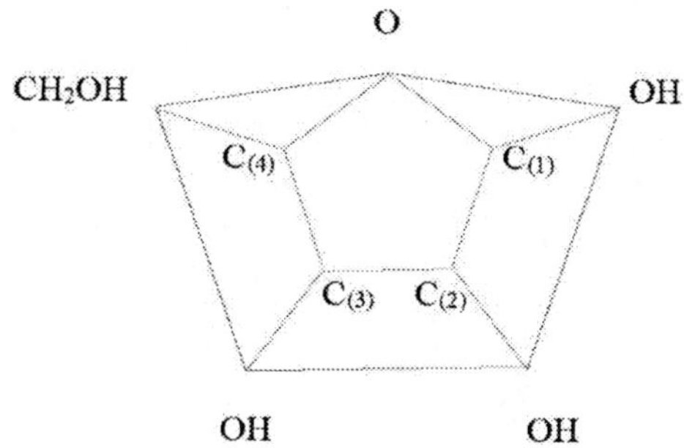

Figure 8.11. The simplified image of the enantiomer A of α-D-ribose molecule.

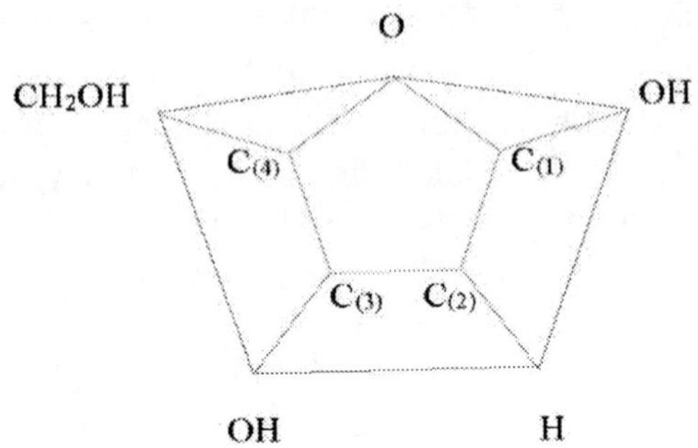

Figure 8.12. The simplified image of the enantiomer A of 2-deoxy-α-D-ribose molecules.

The corresponding images of the α-*D*-ribose and 2-deoxy-α-D-ribose molecules of the enantiomer form *B* are shown in Figure 8.13, Figure 8.14.

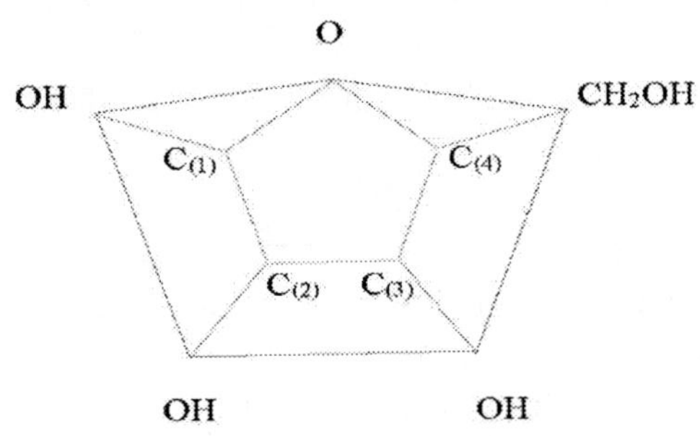

Figure 8.13. The simplified image of the enantiomer B of α-D-ribose molecule.

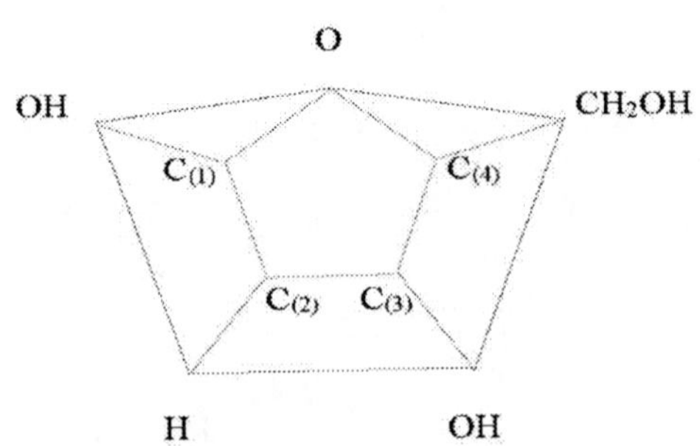

Figure 8.14. The simplified image of the enantiomer B of 2-deoxy-α-D-ribose molecules.

The absence of edges and atoms in Figures 8.11-8.14 in comparison with the Figure 8.10 does not mean that they really are not. This is only the conventionality of the image.

It should be noted that under the assumed projection conditions, the α-*D*-ribose and 2-deoxy-α-*D*-ribose of molecules on a two-dimensional plane in which a closed carbon ring is located has a number of properties that are convenient for further constructions. The enclosed carbon ring is a regular pentagon with an internal angle at the vertices $\pi - \frac{2\pi}{5} = 108^o$. The angles in the three trapezoids formed by the valence bonds and the outer contour, because of the symmetry and parallelism of the bases, are 54 degrees and 126 degrees. In two obtuse-angled isosceles triangles formed by valence bonds and the outer contour, the angles are 126 degrees and 27 degrees.

Now, in a projection onto a two-dimensional plane, one shall depict a sequence of the simplified images of α-*D*-ribose molecules and phosphoric acid residues, taking into account the separation of water molecules at the junction of phosphoric acid residues and α-*D*-ribose molecules and the linear arrangement of the valence bonds in vicinity of these compounds. On Figure 8.15 presented the sequence for α-*D*-ribose molecules in form *A*.

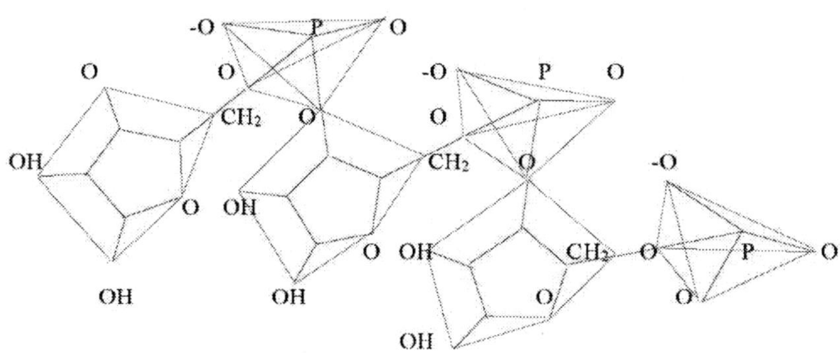

Figure 8.15. Structure of the nucleic acid molecule with α-D-ribose molecules in form A.

On Figure 8.16 presented the sequence structure of the nucleic acid molecule with for *D*-ribose molecules in form *B*.

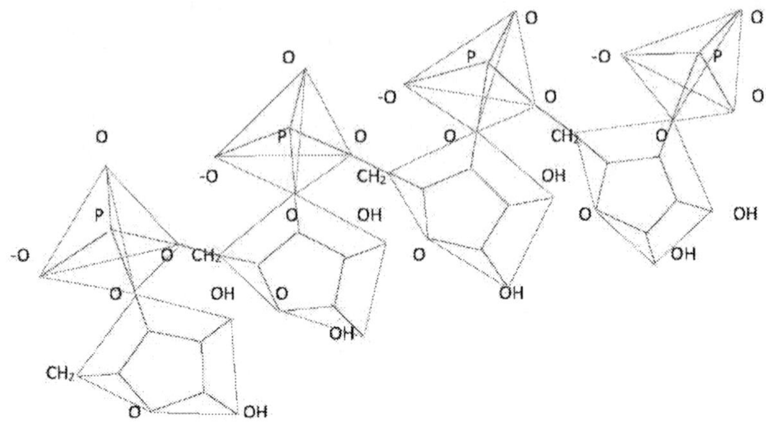

Figure 8.16. Structure of the nucleic acid molecule with D-ribose molecules in form B.

On Figure 8.17 presented the sequence for 2-deoxy-α-*D*-ribose molecules in form *A*.

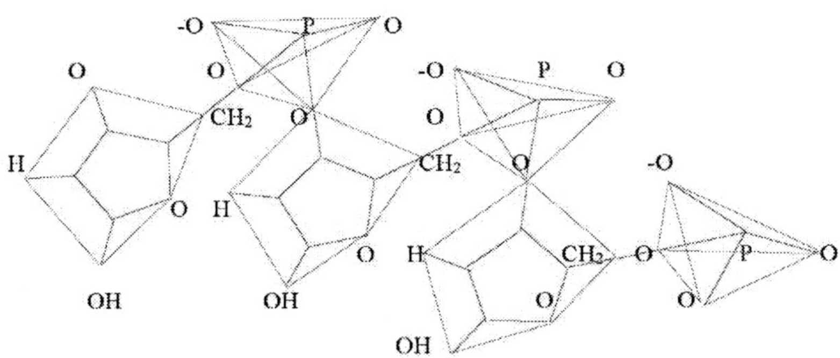

Figure 8.17. Structure of the nucleic acid molecule with 2-deoxy-α-D-ribose molecules in form A.

On Figure 8.18 presented the sequence structure of the nucleic acid molecule with 2-deoxy- *D*-ribose molecules in form *B*.

It follows from Figures 8.15-8.18 that the *D*-ribose molecules in both chains are not linearly distributed. Any sequence of corresponding points of *D*-ribose molecules in chains forms a curved line with rotation to the right side. However, in these sequences there is a fundamental difference. It is a consequence of the use in these two nucleic acid structures of two enantiomeric molecules of α-*D*-ribose: the α-*D*-ribose molecule in form *A* in the chain in Figure 8.15, the α-*D*-ribose molecule in the *B* form in the chain in Figure 8.16. This difference consists in the fact, that in the chain in Figure 8.15 the oxygen atom in the carbon ring of the α-*D*-ribose molecule is the extreme right in each α-*D*-ribose molecule. And in the chain in Figure 8.16 the oxygen atom in the carbon ring of the α-*D*-ribose molecule is the extreme left in the carbon ring of the molecule α-*D*-ribose. It can be argued that the nucleic acid structures in Figures 8.15 and 8.16 are enantiomeric nucleic acid structures, which are a consequence of the presence of two enantiomeric structures of α-*D*-ribose molecules in them. The existence of two enantiomeric nucleic acid structures was not noted before since details of the arrangement of atoms in the nucleic acid structures were not considered. Denote the structure of the nucleic acid in Figure 8.15 of the nucleic acid of the *EA*, and the structure of the nucleic acid in Figure 8.16 of the *EB*.

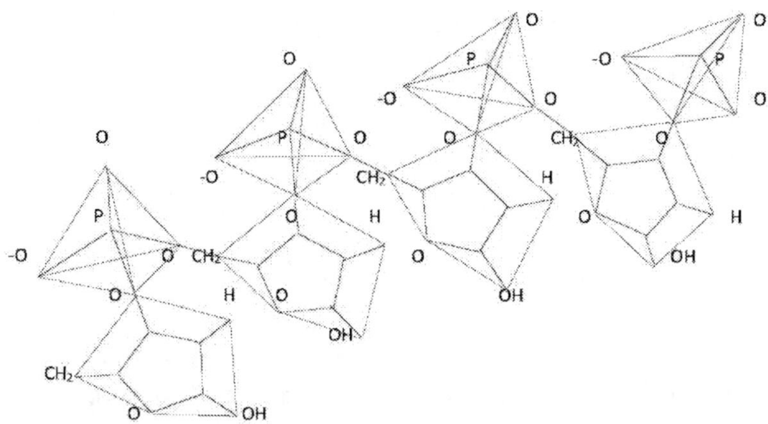

Figure 8.18. Structure of the nucleic acid molecule with 2-deoxy-α-D-ribose molecules in form B.

The projection of the nucleic acid molecules depicted in Figures 8.15 and 8.16 can be described mathematically. This description will also be valid for nucleic acids in which 2-deoxy-*α-D* -ribose molecules are used instead of *α-D*-ribose molecules, since in these cases the only difference between 2-deoxy-*α-D*-ribose molecules and *α-D*-ribose molecules is the replacement at the carbon atom $C_{(2)}$ hydroxyl group per hydrogen atom. This difference does not affect the results of mathematical analysis of the geometry of molecules.

Let us turn to a detailed picture of the connection of two α-*D*-ribose molecules of the form *B* in Figure 8.16 (Figure 8.19).

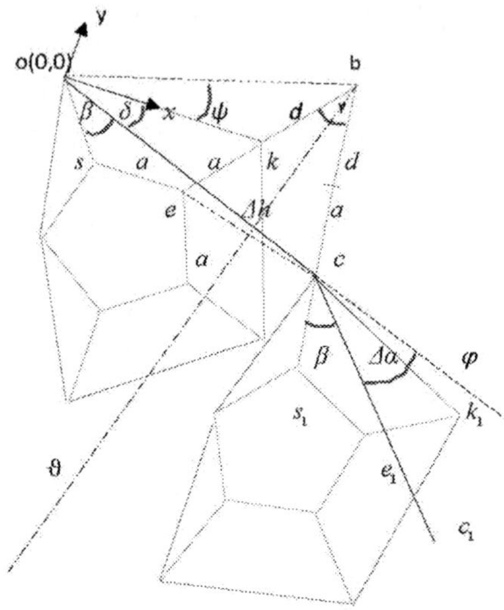

Figure 8.19. Two α-D-ribose molecules of the form B in the structure of the EB.

The vertex *b* in this figure corresponds to the phosphorus atom, and the two edges emanating from it correspond to the projections of two valence bonds of the atom phosphorus with two oxygen atoms. They are at some angle γ on the projection. This angle characterizes the location of the residual phosphoric acid in space with respect to the α-*D*-ribose molecule *B* at the site of their connection (vertex *k*). It can be shown that the angle γ is linearly related to the angle $\Delta\alpha$ between two neighboring nuclei in the spiral chain. Indeed, in Figure 8.19, the smallest distance between two corresponding points of two neighboring α-*D*-ribose molecules *B* in the chain Δh corresponds to the segment *OC*. The smallest angle $\Delta\alpha$ characterizing the helical surface corresponds to the angle between the straight line that is the extension of the *OC* segment and the next segment CC_1 between two molecules of α-*D*-ribose, originating from the vertex *C*. The angle $\Delta\alpha = \angle ocb - \beta$, where $\angle ocb = \angle s_1 c\varphi$ β is the angle between segment *OC* and edge *OS* in the initial molecule α-*D*-ribose. The same angle is repeated in the next molecule of α-*D*-ribose. The angle $\beta = 54^o - \delta$, where 54° is the angle between edge *OS* and edge *OK*. The angle

$\angle obc = \angle obk + \gamma$, but $\angle obk = \pi - \angle okb - \psi, \angle okb = 126^o$ by construction, therefore $\angle obk = 54^o - \psi$.

On the other side it is

$$\angle ocb = \pi - \angle cob - \angle obc = \pi - (\delta + \psi) - \angle obk - \gamma = 126^o - \delta - \gamma.$$

Therefore $\Delta\alpha = 126^o - \delta - \gamma - (54^o - \delta) = \frac{\pi}{5} - \gamma.$

An amazing result was obtained: the rotation angle between two molecules of α-*D*-ribose in the chain depends linearly only on the angle between the projections of the valence bonds in the phosphoric acid residue, that is, the orientation of the phosphoric acid residue in the space with respect to the α-*D*-ribose molecule

$$\Delta\alpha = \frac{\pi}{5} - \gamma. \tag{8.2}$$

Concordantly with model of Watson and Crick (Watson & Crick, 1953 a, b) in period of nuclei chain included 10 nuclei. In this chain the angle $\Delta\alpha = 36^o$. Therefore in this case, according to (8.2), the angle γ is also equal to 36 degrees. If in the period of nuclei chain include 12 nuclei (see Ha, et al., 2005), so $\Delta\alpha = 30^o$, $\gamma = 42^o$.

It is possible to calculate the minimum distance between the corresponding points of neighboring α-*D*-ribose molecules in the nucleic chain Δh . It follows from Figure 8.19 that $\Delta h = \sqrt{y_c^2 + x_c^2}$.

Values y_c, x_c

$$\begin{aligned} 4(d+a)^2 \cos^2 \frac{\pi - \gamma}{2} &= (y_e - y_c)^2 + (x_e - x_c)^2, \\ (d+a)^2 &= (y_b - y_c)^2 + (x_b - x_c)^2, \end{aligned} \tag{8.3}$$

where

$$x_e = a(1 + \cos 54^o), y_e = -a \sin 54^o, x_b = a(1 + 2\cos 54^o) + d \sin 36^o, y_b = d \cos 36^o.$$

The first equation of system (8.3) can obtain from the equality of the square of the length of the segment *ec* to the sum of the squares of the differences of the coordinates of its vertices, on the one hand. On the other hand, the square of the length of the segment *ec* is the square of the doubled product of the cosine of the angle at the base of the isosceles triangle *ecb* by the length of the force side. The second equation of system (8.3) is obtained from the equality of the square of the side length *cb* to square of the sum of the segments *a* and *d*, on the one hand, and to the sum of the squares of the differences of the coordinates of its vertices, on the other hand. Solving system (8.3) for given *a*, *d*, one can find the value of Δh. To find the solution, the squares of the coordinate differences in the right-hand sides of the equations of system (8.3) are opened and the second equation is subtracted from the first. The linear connection of the coordinates of the vertex *c* is obtained. The equation of this connection is substituted into the second equation of system (8.3) and a quadratic equation is obtained for one of the coordinates of vertex *c*. From two roots of the equation, the root corresponding to Figure 8.19 is chosen. Thus, for each value of the angle γ for given the values of *a*, *d* are the coordinates of the vertex *c*.

It is easy to verify that by mirroring the image on Figure 8.19 with respect to the bisector $b\vartheta$ of the angle γ passing through the vertex *b*, one can obtain an image of two coupled nucleotides corresponding to another enantiomeric form of nucleic acid (Figure 8.15). In this way, the calculation of the parameters of the nucleic acid structure according to the system (8.3) corresponds directly to two enantiomeric forms of the nucleic acid structure. In the beginning, in order to find out the properties of the solutions of the system (8.3) corresponding to possible nucleic acid structures, it can numerically study these solutions, arbitrarily varying the value of the angle γ, which is the defining parameter of this system. Note that in this study it confines ourselves to analyzing one chain of a nucleic acid molecule, although, as is known (Watson, Crick, 1953 a, b; Frank-Kamenetskiy, 1988, 2004, 2010), a DNA molecule consists of two or three chains of the DNA molecules connected to each other by hydrogen bonds. Such a connection of chains can be carried out in the future, relying on the obtained regularities of one chain.

Since, the lengths of chemical bonds between carbon-carbon and carbon-oxygen are ≈ 0.15 nm, can will adopt $a = 0.15$ nm. The length of chemical bond phosphor-oxygen can adopt $d = 0.18$ nm. The results of calculating the characteristics of the nucleic acid structure from the equations of system

(8.3) are shown in the graphs (Figure 8.20) and the numerical values of some parameters are given in the Table 8.1.

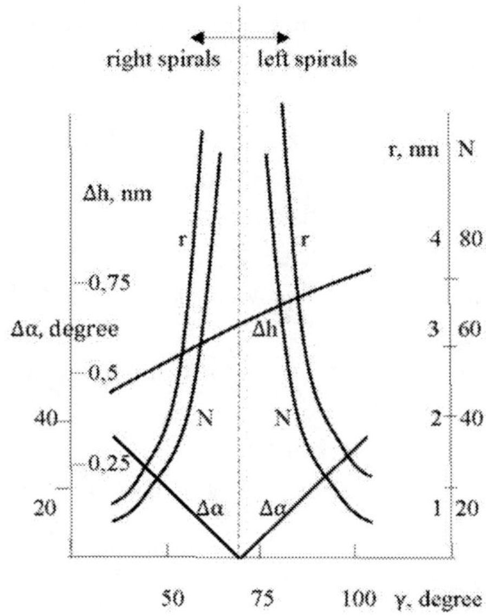

Figure 8.20. Graphs of the change in the distance between nucleotides Δh, the angle of rotation between adjacent nucleotides $\Delta\alpha$, the number of nucleotides in the period N and the radius of the spiral $r = \frac{N\Delta h}{2\pi}$, depending on the angle γ in the projection between the valence bonds of the phosphorus atom.

Table 8.1. The characteristic dimensions of the right and left nucleic acid helices

Γ degree	**Δα degree**	**N**	**Δh Nm**	**R Nm**
36	36	10	0,4677	0,7447
42	30	12	0,501	0,9575
64,5	7,5	48	0,607	4,6397
72	0	∞	0,6394	∞
84	12	30	0,6815	3,2557
90	18	20	0,7	2
102	30	12	0,7337	1,4
108	36	10	0,7639	1,2163

As can be seen from Figure 8.20 and Table 8.1 extension of the angle γ between the projections of the valence bonds of the phosphorus atom in the phosphoric acid residue from the angle of 36 degree to the plane of the carbon ring of the α-*D*-ribose molecule increases almost linearly the distance between the nucleotides in the nucleic acid. In this case, the radius of the nucleic acid helix increases nonlinearly asymptotically to infinity in the neighborhood of the value of the angle γ = 72 degrees. At a value γ of 72 degrees, the nucleic acid molecule is straightened out into a straight line. With further increase in angle γ, the nucleic acid helix begins to rotate to the left. In this range of values, increasing the angle γ also leads to an almost linear increase in the distance between nucleotides, but to a nonlinear decrease in the radius of the nucleic acid helix. This occurs up to an angle γ = 110 degrees, at which the spiral turns into a closed ring, since there is no rise of nucleotides with the help of phosphoric acid residues. In cases where the angle is greater than 72 degrees $\Delta\alpha = \gamma$-72. In Figure 8.21, as an example, an image of two adjacent nucleotides at 90 degrees is presented.

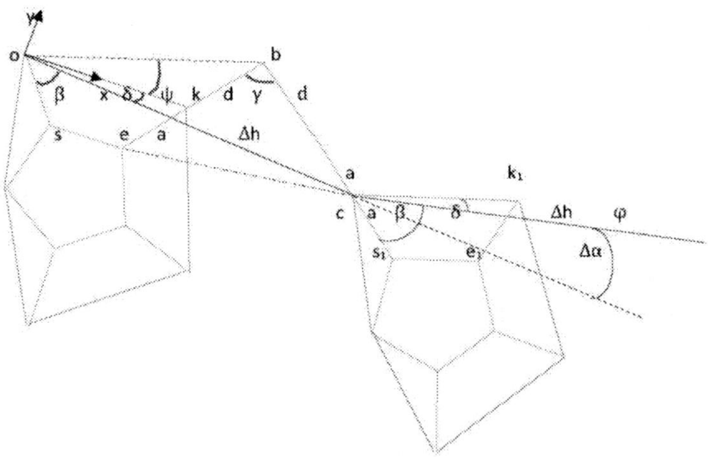

Figure 8.21. Two nuclei of the nucleic acid structure of the EB for γ = 90 degrees.

It can be seen from this figure that when γ is greater than 72 degrees, the angle $\Delta\alpha$ between two adjacent segments is laid to the left of the initial segment. This leads to the rotation of the spiral to the left.

Then from Table 8.1 can see that for γ = 36 degree the distance between nucleotides in structure nucleic acid $\Delta h = 0{,}4677$ nm. Wherein the number of nucleotides in the period is 10 and the radius of the spiral r = 0,7447 nm. To this value of the radius should be added the size of the phosphoric acid residue, which is along the length of the valence bond d and the angle between the valence bonds of 110 degrees $\Delta r = \sin 55^o d2 \approx 0.29$ Then the radius of the nucleic acid helix is $r + \Delta r \approx 1$ nm. The same nucleic acid radius it was observed in experiments (Watson, Crick, 1953 a, b) for 10 of nucleotides in the period. This form of nucleic acid structure was designated B. From Table 8.1can see that for γ = 42 degree the distance between nucleotides in structure nucleic acid Δh =*0,5* nm. Wherein the number of nucleotides in the period is 12 and the radius of the spiral r = 0,957 nm. Then the radius of the nucleic acid helix is $r + \Delta r \approx 1$, 35 nm. These data roughly correspond to the form of the nucleic acid structure, designated A. It is also known from experiments the existence of a nucleic acid structure with 12 nucleotides in a period with left rotation (Ha, et al., 2005; Frank-Kamenetskiy, 2004). This form of nucleic acid structure was designated Z. From Table 8.1 it follows that the value of the angle γ= 102 degrees corresponds to this form. The existence of nucleic acid structures in the form of a closed ring has also been experimentally confirmed (Frank-Kamenetskiy, 2010). This form of nucleic acid structure corresponds to the angle γ= 110 degrees in Table 8.1.

Thus, from the variety of solutions of the system (8.3) corresponding to the possible nucleic acid structures, four forms of the nucleic acid structure were experimentally confirmed. This does not mean that other forms of nucleic acid structure that correspond to other solutions of system (8.3) do not exist. It should be assumed that these solutions have not yet been found in experiments, and in the future, when creating the appropriate conditions, they will be found. The received decisions should be considered also at consideration of forms of structures of nucleic acid consisting of three connected spirals (form H) (Frank-Kamenetskiy, 1988). In addition, must not forget that each solution of the system (8.3) corresponds simultaneously to two enantiomeric forms of the nucleic acid structure.

8.4. The Three-Dimensional Model of the Nucleic Acid Molecule

For clarity, it is convenient to have three-dimensional images of nucleic acids. They will be needed in particular in the next section when analyzing the interaction of nucleic acids. For this purpose, you can use the simplified three-dimensional images of amino acid molecules obtained in Chapter 6. However, it should be remembered here that in reality the molecules of amino acids and nucleic acids have a higher dimension. When analyzing a chain of amino acid molecules, taking into account their higher dimension, in the previous section, we considered the projection of the chemical bond connecting the amino acid molecules with the phosphorus atom of the phosphoric acid residue onto the plane of the oxygen-carbon cycle of the amino acid molecule. Obviously, in this case, the chemical bond itself is located in a space of higher dimension. In the simplified 3D model (Chapter 6), the chemical bond between molecules is orthogonal to the plane of the oxygen-carbon cycle, so its projection onto the plane of the oxygen-carbon cycle is a point. With such a simplification, it would be impossible to obtain the analytical conditions for the existence of various forms of nucleic acids, which were found in the previous section, taking into account the highest dimension of space. Recognizing the importance of the obtained conditions, we nevertheless present a simplified three-dimensional image of the nucleic acid due to its clarity.

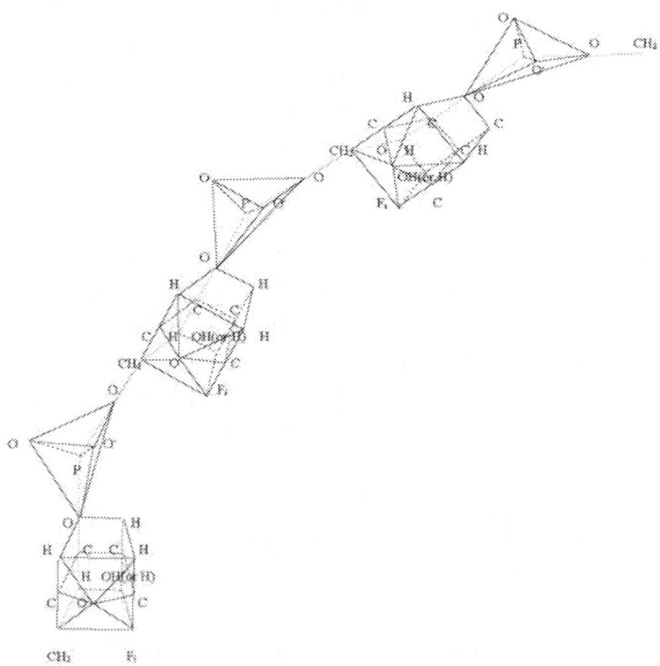

Figure 8.22. A nucleic acid molecule with sugar molecules α-D-ribose (A) or and 2-deoxy-α-D-ribose (A).

A nucleic acid molecule is a chain of residues of phosphoric acid (a tetrahedron with a center) and sugar molecules (a prism with a pyramid). The chain (for molecules α-D-ribose (A) or and 2-deoxy-α-*D*-ribose (A) of successively alternating sugar molecules and residues of phosphoric acid in a three-dimensional form is shown in Figure 8.22.

A nucleic acid molecule with sugar molecules *α-D*-ribose (B) or and 2-deoxy-*α-D*-ribose (B) is shown in Figure 8.23.

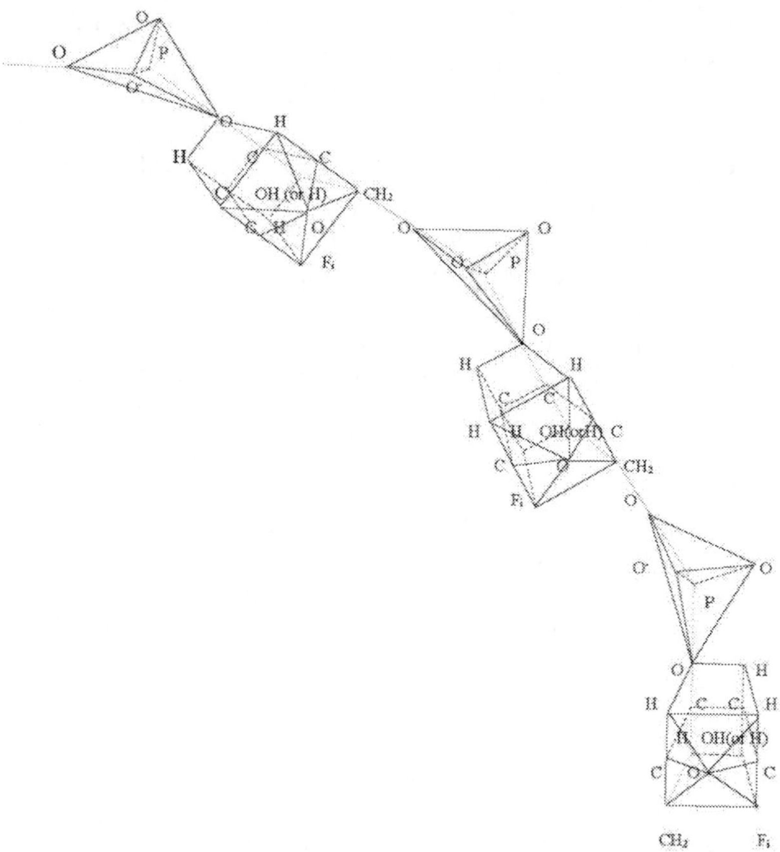

Figure 8.23. A nucleic acid molecule with sugar molecules α-D-ribose (B) or and 2-deoxy-α-D-ribose (B).

At the junctions of the phosphoric acid residue with sugar, due to the separation of water molecules, an oxygen atom and the functional group CH_2 are formed (Figure 8.22, Figure 8.23). In addition, one of the four nitrogenous bases is attached to the $C_{(1)}$ carbon atom instead of the hydroxyl group at the $C_{(1)}$ carbon atom (Figure 8.22, Figure 8.23). The yellow color in Figure 8.22, Figure 8.23 indicates the covalent bond lines connecting the phosphoric acid residues and the sugar molecules.

Figures 8.22 and 8.23 represent two chiral forms of nucleic acids (two chiral forms of DNA and two chiral forms of RNA), since not only five-carbon sugar molecules participate in them in specular reflections, but also phosphoric acid residues are also in specular reflections. Let us consider in more detail one of these forms, presented, for example, in Figure 8.22. It is significant that the communication lines connecting the tetrahedron with the sugar molecule form, by repulsing the electron pairs, uniform straight segments perpendicular to the bases of the prism. In the center of the tetrahedron, the chemical bond lines form a kink, since in the center of the tetrahedron there are straight lines connecting the center with the vertices of the tetrahedron. Ultimately, this fracture with simultaneous movement along the vertical coordinate leads to the formation of a helix of a nucleic acid molecule. In the projection on the plane of the figure, the covalent bond segments form a broken line close to the circle (polygon). The radius of the circle describing this polygon will be determined by the angle between the projections on the plane of the chemical bonds of the phosphorus atom with the oxygen atoms at its vertices belonging to the polygon. The tetrahedron with the center (the residue of phosphoric acid), connecting with the top of the base of the prism, has a degree of freedom. It can be rotated relative to the connection between the vertex of the prism and the center of the tetrahedron at an arbitrary angle. Therefore, even in the projection on the plane perpendicularity to axes of rotation, the angle between the chemical bonds emanating from the center of the tetrahedron can be arbitrary (Figure 8.24).

Let this angle be γ. Then the broken line between the centers of the tetrahedrons closes when the product of the number of these rotations n and the angle π-γ is equal to 2π. (On projection in Figure 8.24 edge of low bases of prism in visible due to the perpendicularity of the generators and bases.) You can find the period of the spiral, given that there is still movement along a line perpendicular to the projection plane. Thus, to determine the radius of a circle, we have the equality

$$(\pi - \gamma)\mathrm{n} = 2\pi.$$

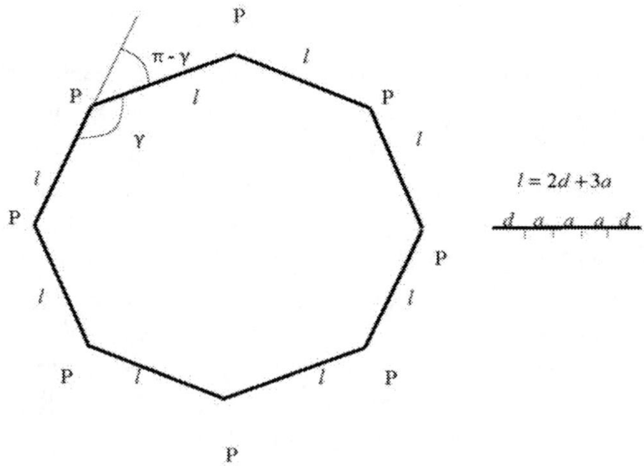

Figure 8.24. The projection of spiral on the plane perpendicularity to axes of rotation.

Since the number of turns equals the number of segments that make up a polygon, one have approximately the equality

$$\frac{2\pi}{\pi-\gamma}=\frac{2\pi R}{l},$$

R is radius of the circle, and l is length of the segment between centers of tetrahedrons.

From the last equality can find the radius of the circle

$$R=\frac{l}{\pi-\gamma}.$$

From Figure 8.22 and Figure 8.24 it follows that $l=3a+2d.$ to the accepted values a and d can get

$R=\dfrac{0.81}{\pi-\gamma}$ nm. For example, in case $\gamma=140^{o}$ is R=1.16 nm.

This radius value is close to the experimental helix radius of nucleic acids measured by Watson and Crick. At this angle between the chemical bonds in the projection of the phosphoric acid residue, the number of nucleotides in the period $\frac{2\pi}{\pi-\gamma}=9.$ The number is in satisfactory agreement with the experimental number of nucleotides 10, given that the helix of the helix in the period is somewhat greater than the circumference of the same radius. This confirms the correctness of the three-dimensional model of the nucleic acid molecule.

The helix surface is formed due to the fact, that in a tetrahedron with a center the valence bonds connecting the center with the vertices do not lie in the same plane (Zhizhin, 2019b). Therefore, the valence bonds connecting the two sugar molecules do not lie together in the plane passing through the edge of the prism. The bond leaving the phosphorus atom, approximately 23 degrees, retreats from the plane in which the bond lies, which is part of the phosphorus atom. This causes a chain to shift along a coordinate perpendicular to the circle. The equation of the helix surface in this case can be written as $h = n\Delta h, \Delta h = l\tan(\Delta\alpha), \Delta\alpha$

previous segment *l* and the subsequent segment *l*, *n* is the number of turns, *h* is the height of the period. Thus, $\Delta h = 0.343nm, h = 3.43nm.$

height value is close to the value found in the Watson and Crick experiments (3.46 nm).

If the valence bond in the tetrahedron connecting its center with the next sugar molecule is rejected in the projection to the left of the valence bond connecting the center of the tetrahedron with the previous sugar molecule, i.e., it is $\gamma > 180^o$ the nucleic acid is a left-handed helix. If is $\gamma = 180^o$, then the nucleic acid has a linear appearance.

In ribonucleic acids, a helix return in one molecule is observed (Spirin, 2019). This is possible if the angle γ is variable, i.e., in the chain, the angle γ increases to 180 degrees and the achievement of even larger values of this angle. This means that along the chain there is a sequential rotation of the tetrahedron around the valence bond connecting its center with the previous sugar molecule.

Obviously, a similar consideration can be made for the single-stranded nucleic acid molecule shown in Figure 8.23. The difference between Figure 8.22 and Figure 8.23 is that the chain helix in Figure 8.22 rotates to the right, the chain helix in Figure 8.23 rotates to the left. In this case, the residues of phosphoric acid at the corresponding positions in the chains are mirrored to each other. Since the phosphoric acid residue has a degree of freedom (rotation around the axis connecting it to the five-carbon sugar molecule), with a different initial state of the phosphoric acid residue in Figure 8.22, there will be another state of the phosphoric acid residue in Figure 8.23. This leads to the set of single-stranded nucleic acids (both DNA and RNA) is broken into two sets of chiral forms. Each form in one set contains a chiral form in another set. Moreover, in each set there are possible rotation of the spirals both in the right and in the left direction.

Conclusion

The currently known forms of the helical structure of nucleic acid molecules *A, B, Z,* and nucleic acid molecules in the form of a closed ring were experimentally found in the works of Watson, J.D., Crick, F.H.C., Ha, S.C., Frank-Kamenetskiy, M. and other scientists. However, the reasons for the formation of such a different form of nucleic acid molecules have not been clarified. In this chapter is proved that the main reason for the formation of nucleic acid molecules of different shapes is the different geometric realization of the chemical bond between the phosphoric acid residue and the *D*-ribose molecule, as objects of higher dimensionality. In other words, the residues of phosphoric acid can geometrically act as a bridge, connecting the molecules of *D*-ribose with each other. It is shown that the angle γ between the projections of the valence bonds of the phosphorus atom in the phosphoric acid residue on the plane of the carbon ring in the *D*-ribose molecule is a parameter characterizing the various geometric forms of chemical bonding of phosphoric acid residues with *D*-ribose molecules. The angle γ characterizes the orientation of the phosphoric acid residue in the three-dimensional space. This parameter can therefore be regarded as an internal degree of freedom of the structure of the nucleic acid molecule.

A mathematical model describing the geometric features of the chemical bond of these objects is constructed. Calculations on this model showed that the characteristics of known nucleic acid structures, determined experimentally, and calculated from the constructed model, are in good agreement with each other, while the model does not contain any empirical parameters. Calculations show that known nucleic acid structures are particular cases in the model and it can be considered that known examples of nucleic acid structure forms are experimental confirmation of the correctness of the mathematical model. However, according to the mathematical model, its solutions are many other nucleic acid structures that have not been experimentally discovered so far. One can hope that their experimental discovery is ahead.

The existence of two enantiomeric forms of nucleic acid molecules, connected with the existence of two enantiomeric forms of *D*-ribose molecules, is established.

The resulting variety of nucleic acid molecule structures is represented in the form of analytical dependencies, graphs and table.

They should be taken into account when analyzing the synthesis of proteins, replication and translation of nucleic acid in the work of ribosomes (Stewart & McLachlan, 1975; Spirin & Gavrilova, 1971; Spirin, 1986).

The higher dimensionality of the constituent nucleic acid molecules, which allows to describe mathematically the structure of nucleic acid, requires reconsidering the issues of tight packing of nucleic acid molecules in cells, viruses and bacteria, provided that the nucleic acid chains necessary for the preservation and transfer of genetic information are complementary.

Chapter 9

Interaction of Nucleic Acids in the Space of Higher Dimension and the Transmission of Hereditary Information

Abstract

The geometry of the neighborhood of the compound of two nucleic acid helices with nitrogen bases has been investigated in detail. It is proved that this neighborhood is a cross-polytope of dimension 13 (polytope of hereditary information), in the coordinated planes of which there are complementary hydrogen bonds of nitrogenous bases. It was found that methylation of viral nucleic acids leads to the transition of the nitrogenous bases into the space of the higher dimension and sharply (more than 2000 times) the intensity of information processes in the field of interaction of nucleic acids (in the polytope of hereditary information).

Keywords: dimension of the space, genes, hereditary, incidence coefficient, n-cross-polytope, virus

Introduction

There are two known nucleic acids: deoxyribonucleic acid (DNA) and ribonucleic acid (RNA). Direct experiments by Avery, McLeod and McCarthy (1944) proved that DNA can be the carrier of the hereditary traits of an organism, i.e., carrier of genetic information. At the same time, RNA is an intermediary that receives this information from DNA and implements it in the process of protein biosynthesis. Watson and Crick (1953) proposed a model of the macromolecular structure of DNA. This is a double helix, where two DNA polymer strands are twisted relative to each other around a common axis and are held together due to pair interactions of nitrogenous bases covalently bonded to five-carbon sugar molecules. In turn, sugar molecules are linked to each other by phosphoric acid residues, forming a chain. Bases are large circular molecules with nitrogen atoms. DNA nucleotides (a sugar molecule, a phosphoric acid residue and a base) have one of the four bases adenine (A), guanine (G), cytosine (C) and thymine (T). In RNA, thymine is replaced by uracil (U). Cytosine, thymine and uracil have one ring of atoms and are called pyrimidine bases. Adenine and guanine each have two rings and are called purine bases. It is significant that nitrogenous bases are practically flat. The reproduction (reduplication) of the DNA structure is based on the principle of complementarity: in a double helix, two DNA polymer chains are linked to each other by hydrogen bonds in the formation of pairs G: C, C: G, A: T, T: A (the colon denotes a hydrogen bond). If two chains of the double helix diverge, then a new complementary chain can be built on each of them. At the same time, opposite the G of the original chain, C of the new chain is established, opposite C of the old chain-G of the new chain, opposite A-T, and opposite T-A. As a result, two child double helixes are obtained, completely identical to the original parent. The same principle of complementarity provides a mechanism for RNA replication on a DNA template. This is another type of interaction between nucleic acids. The only difference is that RNA polymerizes only on one of the two separated strands of the DNA double helix. During RNA synthesis, opposite A of the DNA chain becomes U (uridyl ribonucleide), instead of T during DNA synthesis. The replicating strand of RNA is thus an exact copy of the opposite strand of DNA, with T replaced by Y. As a result of this replication, RNA is formed as a flexible single-stranded polymer.

By the second half of the 1950s, it was established that protein synthesis in living cells is carried out by ribonucleoprotein particles-ribosomes. The main component of ribosomes is ribosomal RNA, it consists of transfer RNA that transports selected amino acids to messenger RNA, which is synthesized and edited as a result of RNA processing. This is the third type of nucleic acid interaction. It was found that due to the interaction of side groups (nitrogenous bases) with each other, a single-stranded RNA polymer folds onto itself, forming secondary and tertiary structures (Spirin, 2009). The formation of these structures also occurs due to complementary interactions result of the interaction of nucleic acids, the central dogma of molecular biology is fulfilled: the flow of hereditary information is transmitted from DNA to RNA and further to protein. At present, in all types of interaction of nucleic acids, it is believed that nitrogenous bases are linearly sequential along the polymeric chain of acids. But since there are four traditional types of nitrogenous bases in each chain, and there are 20 types of amino acids in proteins, it has come to be recognized that each amino acid corresponds to three nitrogenous bases. This assumption is based on the concept of the genetic code, the deciphering of which establishes these correspondences (Gamov, 1954; Crick, 1959). However, in recent works (Zhizhin, 2019, 2020 a, b; 2021) it was found that the region of interaction of nucleic acids is not a linear section of the chain, but a polytope of higher dimension (a polytope of hereditary information), at the tops of which nitrogenous bases are located. In this regard, the question of the interaction of nucleic acids requires additional consideration. This chapter is dedicated to this.

9.1. Polytopes with Antiparallel Edges

In single-stranded and double-stranded nucleic acids (RNA, DNA), the constituents of acids (residues of phosphoric acid and sugar molecules) interact with each other (Watson & Crick, 1953a, b; Spirin & Gavrilova, 1971; Frank-Kamenetskiy, 1986, 1988). Phosphoric acid residues are connected by divalent metal ions, mainly magnesium ions, due to the interaction of negative charges of phosphoric acid residues with positive charges ions (Spirin & Gavrilova, 1971). This interaction is essential for the stability of nucleic acid structures, especially in the ribosomes. Sugar molecules interact with each other due to the hydrogen bond between the nitrogenous bases attached to the sugar molecules. The constituents of nucleic acids being geometric forms interact with each other to form new geometric forms-new polytopes. However, it is not known how flat nitrogenous bases are oriented in space, whether their orientation depends on the type of nitrogenous base. Currently there is no information on this. There is also no information on how exactly the metal ions are located, connecting the phosphoric acid residues. It should be remembered here that the adopted three-dimensional model of the components and the nucleic acid molecule itself is only a model for visual perception. As it was shown earlier, the phosphoric acid residue is a polytope of dimension 4, and the sugar molecule has a dimension of 12.

The movement of triangles along a helix leads to the formation of polytopes with antiparallel edges. Consider an arbitrary triangle *ABC* on the plane. Choose some point O' on the plane outside the triangle to his left. Let this point be the base of the axis of the helix passing through the triangle. Rotate the *ABC* triangle 180 degrees to the right, moving it up the helix, parallel to the initial plane. In the projection on the plane, both triangles ABC and the displaced triangle $A'B'C'$

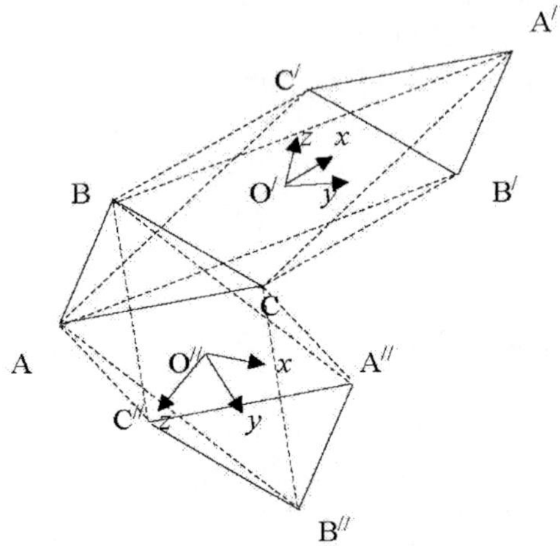

Figure 9.1. Polytopes of dimension 3 with anti-parallel edges.

It is easy to see that the edges of the triangle ABC and $A'B'C'$ antiparallel. It can now connect in space the vertices of the triangle ABC with the vertices of the triangle $A'B'C'$
vertices with the same letters. In a projection on the plane the connections are represented by dotted segments. It can be seen that the connecting segments also break up into pairs of antiparallel segments. Let us now verify that the image $ABCA'B'C'$ in Figure 9.1, along with the dotted segments, is a projection of a three-dimensional convex polytope. We use the Euler-Poincaré equation (Poincaré, 1895) for this aim

$$\sum_{k=0}^{n}(-1)^k f_k(n) = (-1)^k + 1, \tag{9.1}$$

f_k is the number of elements of dimension k in polytope of dimension n.

The shape $ABCA'B'C'$ in Figure 9.1 has 6 vertices, 12 edges, 8 triangular faces (rectangles are not faces by construction, since connecting, for example, vertex A with vertices B', C'

triangle $AB^{/}C^{/}$

dimensions into equation (9.1), can find 6-12 +8 = 2, i.e., the Euler-Poincaré equation holds in this case for $n = 3$. This proves that the resulting figure is a convex polytope of dimension 3 (if the figure were not convex, the Euler-Poincaré equation would be violated). The point $O^{/}$ in Figure 9.1 coincides with the center of the three-dimensional figure $ABCA^{/}B^{/}C^{/}$ as diagonal figures pass through it and point $O^{/}$ located on the axis of the helix. Point $O^{/}$ can be considered as the origin of three-dimensional space. Coordinate axes *x, y, z* in this direction emanates from directions $AA^{/}, BB^{/}, CC^{/}$

Three pairs of these axes define the coordinate planes of the space of this shape. Choose some point $O^{//}$ on the plane outside the triangle below it. Let this point be the base of the axis of the helix passing through the triangle. Rotate the *ABC* triangle 180 degrees to the left, moving it up the helix, parallel to the initial plane. In the projection on the plane, both triangles *ABC* and the displaced triangle $A^{//}B^{//}C^{//}$

It is easy to see that the edges of the triangle *ABC* and $A^{//}B^{//}C^{//}$ antiparallel. It can now connect in space the vertices of the triangle *ABC* with the vertices of the triangle $A^{//}B^{//}C^{//}$

vertices with the same letters. In a projection on the plane the connection are represented by dotted segments. It can be seen that the connecting segments also break up into pairs of antiparallel segments. Let us now verify that the image $ABCA^{/}B^{/}C^{/}$ in Figure 9.1, along with the dotted segments, is a projection of a three-dimensional convex polytope. Obviously, a shape $ABCA^{//}B^{//}C^{//}$ has as many vertices, edges, and flat faces as a shape $ABCA^{/}B^{/}C^{/}$. Therefore, it satisfies the Euler-Poincaré equation (9.1) with dimensionality $n = 3$, i.e., it is a convex three-dimensional polyhedron. The point $O^{//}$ in Figure 9.1 coincides with the center of the three-dimensional figure $ABCA^{//}B^{//}C^{//}$ as diagonal figures pass through it and point $O^{//}$ located on the axis of the helix. Point $O^{/}$ can be considered as the origin of three-dimensional space. Coordinate axes *x, y, z* in this direction emanates from directions $AA^{//}, BB^{//}, CC^{//}$

x, y, z in the figures $ABCA^{/}B^{/}C^{/}$ and $ABCA^{//}B^{//}C^{//}$, as can be seen from Figure 9.1, coincide. This suggests that both figures $ABCA^{/}B^{/}C^{/}$ and $ABCA^{//}B^{//}C^{//}$ are topologically one and same figure-the wrong octahedron.

Interestingly, to transform an arbitrary tetrahedron $ABCD$ into a tetrahedron $A^{/}B^{/}C^{/}D^{/}$

along a helix by 180 degrees. To do this, you must turn the helix together with the tetrahedron and move the tetrahedron along the helix in the opposite direction, also rotating it 180 degrees. In the initial state, the tetrahedron on the initial helix and the tetrahedron on the reversed helix, after its rotation by 180 degrees, will have anti -parallel edges. Both tetrahedrons can *ABCD* and $A'B'C'D'$ be present by the Figure 9.2 for rotation to the right. On Figure 9.2 the point O' is the point of rotation.

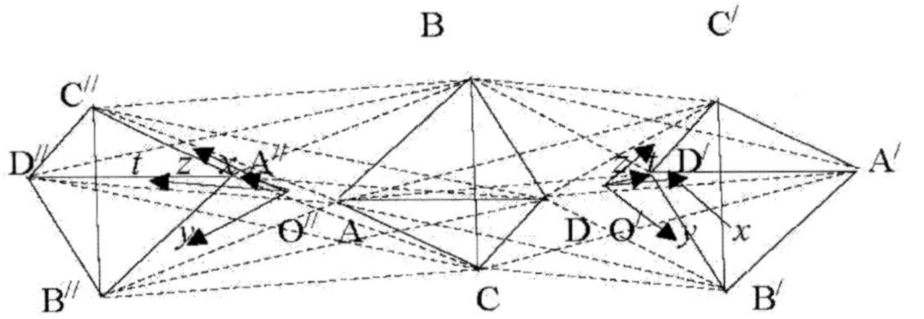

Figure 9.2. Polytope of dimension 4 with anti-parallel edges.

Now connect the vertices of the tetrahedrons so that the connecting edges (dotted segments) do not have the same letters. The resulting figure (along with dotted edges) has 8 vertices (f_0 = 8), 24 edges (f_1

triangular faces (f_2 = 24), and 8 tetrahedrons (f_3

values into equation (9.1), can find 8-24 + 24-8 = 0. Consequently, the Euler-Poincaré equation is satisfied in this case for n = 4. Thus, the polytope $ABCDA'B'C'D'$ in Figure 9.2 has dimension 4. It is easy to see (Zhizhin, 2019 b) that this is 4-cross-polytope (Grunbaum, 1967; Zhizhin, 2014). The point O' in Figure 9.2 coincides with the center of the fourth-dimensional figure $ABCDA'B'C'D'$ as diagonal figures pass through it and point O' located on the axis of the helix. Point O' can be considered as the origin of fourth-dimensional space. Coordinate axes *x, y, z, t* in this direction emanates from directions AA', BB', CC', DD' respectively. Six pairs of these axes define the coordinate planes of the space of this shape.

Choose some point O'' on the plane outside the tetrahedron *ABCD* below it. Let this point be the base of the axis of the helix passing through the tetrahedron. Rotate the *ABCD* tetrahedron 180 degrees to the left, moving it up the helix, parallel to the initial plane. In the projection on the plane, both tetrahedrons *ABCD* and the displaced tetrahedron $A''B''C''D''$

located as shown in Figure 9.2. It is easy to see that the edges of the tetrahedrons *ABCD* and $A^{//}B^{//}C^{//}D^{//}$
space the vertices of the tetrahedron *ABCD* with the vertices of the tetrahedron $A^{//}B^{//}C^{//}D^{//}$
with the same letters. In a projection on the plane the connection is represented by dotted segments. It can be seen that the connecting segments also break up into pairs of antiparallel segments. Let us now verify that the image $ABCD\,A^{//}B^{//}C^{//}D^{//}$ in Figure 9.2, along with the dotted segments, is a projection of a fourth-dimensional convex polytope. Obviously, a shape $ABCD\,A^{//}B^{//}C^{//}D^{//}$ has as many vertices, edges, and flat faces as a shape $ABCD\,A^{/}B^{/}C^{/}D^{/}$. Therefore, it satisfies the Euler-Poincaré equation (9.1) with dimensionality $n = 4$. Thus, the polytope $ABCD\,A^{//}B^{//}C^{//}D^{//}$ in Figure 9.2 has dimension 4. It is easy to see (Zizhin, 2019 b) that this is 4-cross-polytope.

The point $O^{/}$ in Figure 9.2 coincides with the center of the 4-cross-polytope as diagonal figures pass through it and point $O^{//}$ located on the axis of the helix. Point $O^{//}$ can be considered as the origin of fourth-dimensional space. Coordinate axes *x, y, z, t* in this direction emanates from directions $AA^{//}, BB^{//}, CC^{//}, DD^{//}$
that the sequence of alternation of coordinate axes *x, y, z, t* in a 4-cross-polytope $ABCD\,A^{/}B^{/}C^{/}D^{/}$ differs from the sequence of alternation of coordinate axes *x, y, z, t* in a 4-cross-polytope $ABCD\,A^{//}B^{//}C^{//}D^{//}$. Thus, a surprising fact emerged: the figures $ABCD\,A^{/}B^{/}C^{/}D^{/}$ and $ABCD\,A^{//}B^{//}C^{//}D^{//}$, being 4-cross-polytopes, are topologically different from each other. It is impossible to move from one of them to another by continuous transformation since they have a different order of alternation of vertices.

9.2. The Polytope of Hereditary Information

Let us consider in detail the formation of a polytope of two sugar molecules with anti-parallel edges. Here, as in the case of the tetrahedron, to form a polytope with antiparallel edges from two sugar molecules, you must have one sugar molecule on one helix to turn this helix together with the sugar molecule and move the sugar molecule along this reversed helix in opposite direction to the original helix direction. When the sugar molecule rotates 180 degrees to the right while moving, then the original sugar molecule and the sugar molecule on the reversed helix are two polytopes with anti-parallel edges. Both of these sugar molecules in a simplified form are represented in Figure 9.3.

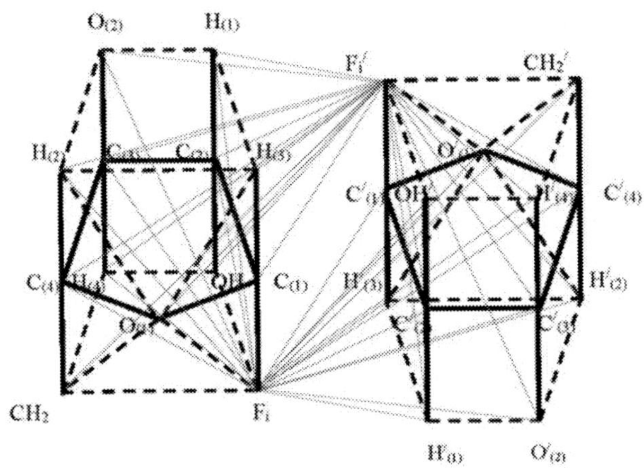

Figure 9.3. The two sugar molecules in a simplified form with anti-parallel edges.

When the five-carbon sugar molecule is rotated to the left by 180 degrees, the nitrogenous bases are on opposite sides of both molecules, so that for their connection it is necessary to cross the entire set of atoms of two molecules. This is unrealistic therefore this option is not considered. When the sugar molecule is rotated 180 degrees to the left, it is possible to connect the sugar molecules through nitrogenous bases only between two different chains of nucleic acids or in the case of turning the chain itself in the opposite direction.

In full, the sugar molecules have a dimension of 12, in the corresponding polytope each vertex must have an edge connection with all the other vertices. Knowledge of this now one need. For the formation of a polytope of dimension 13, it is necessary to connect each vertex of one polytope with the vertices of another polytope so that there are no vertex connections with the same letters. All connecting edges break into pairs of antiparallel edges. At the same time, a set of coordinate a two-dimensional planes emanates from the center of the formed polytope as from the origin of coordinates. Their number is equal to the number of combinations from 13 to 2, i.e., 48 coordinate planes. In order to clarify the possible geometrical circumstances of the connection of helices in double-helix nucleic acid molecules with nitrogenous bases, we are primarily interested in the coordinate planes containing these nitrogenous bases $F_i, F_i^/$. There are exactly 12 such coordinate planes in the obtained polytope of dimension 13. They are depicted in Figure 9.3 by blue solid lines and are indicated below by the vertices of the polytopes contained in them

$$\frac{F_i^/ H_{(1)}}{H_{(1)}^/ F_i}, \frac{F_i^/ O_{(2)}}{O_{(2)}^/ F_i}, \frac{F_i^/ H_{(2)}}{H_{(2)}^/ F_i}, \frac{F_i^/ C_{(3)}}{C_{(3)}^/ F_i}, \frac{F_i^/ C_{(2)}}{C_{(2)}^/ F_i}, \frac{F_i^/ H_{(3)}}{H_{(3)}^/ F_i},$$

$$\frac{F_i^/ C_{(4)}}{C_{(4)}^/ F_i}, \frac{F_i^/ H_{(4)}}{H_{(4)}^/ F_i}, \frac{F_i^/ OH}{(OH)^/ F_i}, \frac{F_i^/ C_{(1)}}{C_{(1)}^/ F_i}, \frac{F_i^/ O_{(1)}}{O_{(1)}^/ F_i}, \frac{F_i^/ CH_2}{(CH_2)^/ F_i}$$

Other edges of the polytope of dimension 13 are not shown in Figure 9.3, so as not to ignite the figure. In the center of each parallelogram, indicated by its four vertices, is the origin of coordinates and the corresponding pair of coordinate axes (they are not shown). To identify the different hydrogen and oxygen atoms at vertices of polytope, they indicated by numbers in brackets at the lower indices.

It is surprising that the number of coordinate planes containing vertices is exactly equal to the number of possible compounds of nitrogenous bases 12 (Spirin, 2019)

$$A:U, U:A, G:C, C:G, G:U, G:A, U:U, U:C, A\cdot A, A\cdot C, U:G, A:G.$$

Since each coordinate plane designated by the vertices of the parallelograms has a specific atomic environment, it can be assumed that each of the 12 possible compounds of nitrogenous bases is located on one particular coordinate plane out of 12 possible. This solves the question of the possible orientation of the bond of nitrogenous flat bases in nucleic acids using ideas about the high dimensionality of the constituent nucleic acids. It is also surprising that in order to create 13-cross-polytopes, providing the connection with nitrogenous bases $F_i, F_i^/$, nature specially created double-stranded nucleic acids with oppositely directed spirals. This is realized in DNA and RNA when creating regions with inverted helices.

In a variety of nucleic acid molecules, the issue of chain interaction is important. In ribosomes, RNA interacts with each other due to bivalent metal ions (mainly magnesium ions). Positively charged magnesium ions attract negative charges of phosphoric acid residues, ensuring the stability of the ribosomes. In double-stranded nucleic acids, the helices are connected to each other by means of nitrogenous bases complementarily interacting with each other by a hydrogen bond. However, the magnesium ions and nitrogen bases in nucleic acids could not be specifically located. It has been established that magnesium ions and flat nitrogenous bases can be located inside special polytopes of higher dimension. Here knowledge is needed of the higher dimension of phosphoric acid residues and sugar molecules. Such polytopes are polytopes with anti-parallel edges, i.e., cross-polytopes of higher dimension. Binding agents are located on the free coordinate planes of these polytopes in the vicinity the center of the polytope.

In this case, the two-dimensional coordinate plane on the boundary of the polytope should contain the objects to be joined. In the case of magnesium ions, there are four specific coordinate planes inside the 5-cross-polytope, in which an ion can accommodate, combining negative charges. In the case of nitrogenous bases, the existence of 12 coordinate planes inside a cross-polytope of dimension 13, in which flat nitrogenous bases can be located, connecting the helix of nucleic acids, was discovered. Exactly as much as there are options for combining nitrogenous bases. It is given, that each coordinate plane of these 12 planes has a specific environment of atoms. It should be assumed that only one of the 12 possible compounds of nitrogenous bases is placed in each of these planes. It is surprising that the existence of higher-dimensional polytopes with anti-parallel edges is possible only in the case of the opposite direction of interacting helices, and this is exactly what nature provides in the double-helix DNA and in the RNA segments with self-inversion of the helix in the opposite direction.

To build the polytopic of hereditary information, we need to supplement Figure 9.3 with edges that create a closed and convex figure. In this case, it is not necessary to connect the vertices symmetrically relative to the center of Figure 9.3 with the edges. Each vertex will be connected by an edge to the other vertices. In this case, the polytope will be a cross-polytope and its dimension is equal to half the number of vertices. Indeed, according to (Zhizhin, 2019 b), there is a relation between f_i

elements i in the cross-polytope and the dimension d of the cross-polytope itself

$$f_i(d) = 2^{1+i} C_d^{d-1-i} \tag{9.2}$$

At the vertex $i = 0$, therefore, according to (9.2) $f_0 = 2C_d^{d-1} = 2d, d = f_0 / 2 = 13$.

Thus, the polytope of the hereditary information has dimension 13. When portraying this polytope, we will use the technique that is used in portraying cross-polytopes. We distribute all 26 vertices on the circle so that the vertices opposite in Figure 9.3 remain opposite and there is no edge between them (Figure 9.4).

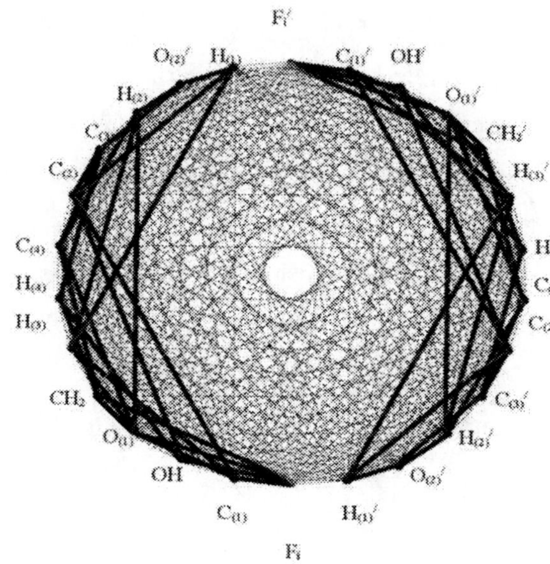

Figure 9.4. The polytope hereditary information.

The edges corresponding to the chemical bonds are indicated in this figure by thick black solid lines. Anti-parallel edges highlighting coordinate planes with vertices corresponding to nitrogenous bases are indicated by bluey lines. The remaining edges are indicated by thin dash-dotted black lines. We emphasize that there is a chemical bond between the nitrogenous bases, but it is not covalent. Therefore, this connection is not indicated by an edge.

When the polytope of hereditary information moves along the common axis of the double helix of nucleic acids in accordance with the sequence of bases in DNA, the connection between the vertices $F_i^{/}, F_i^{//}$
one of the twelve base pairs, which occupies one of the 12 coordinate planes of the polytope.

By relation (9.2), one can determine the number of elements of different dimensions in the polytope of hereditary information. The number of edges is $f_1(13) = 2^2 C_{13}^{11} = 312$
$f_2(13) = 2^3 C_{13}^{10} = 2288$, the number of tetrahedrons is $f_3(13) = 2^4 C_{13}^{9} = 10560$, the number of fourth-dimensional faces (simplexes) is $f_4(13) = 2^5 C_{13}^{8} = 41184$
faces (simplexes) is $f_5(13) = 2^6 C_{13}^{7} = 109824$
dimensional faces (simplexes) is $f_6(13) = 2^7 C_{13}^{6} = 219648$
seven-dimensional faces (simplexes) is $f_7(13) = 2^8 C_{13}^{5} = 164736$
number of eight-dimensional faces (simplexes) is $f_8(13) = 2^9 C_{13}^{4} = 183040$
the number of nine-dimensional faces (simplexes) is $f_9(13) = 2^{10} C_{13}^{3} = 292864$, the number of ten-dimensional faces (simplexes) is $f_{10}(13) = 2^{11} C_{13}^{2} = 159744$
dimensional faces (simplexes) is $f_{11}(13) = 2^{12} C_{13}^{1} = 26624$
twelve-dimensional faces (simplexes) is $f_{12}(13) = 2^{13} = 4096$.

The obtained numbers determine the structure of the polytope of hereditary information.

9.3. Hidden Nucliec Acid Bond Order

From the two sugar molecules in a simplified form Figure 9.6 it follows, however, that not all 12 coordinate planes can have flat nitrogenous bases. They could be located only between the vertices of the polytope belonging to different sugar molecules closest to each other, i.e., between the planes

$OH\text{-}\ H_{(3)}\text{-}H_{(1)}\text{-}F_i \qquad OH^{/}\text{-}\ H_{(3)}{}^{/}\text{-}H_{(1)}{}^{/}\text{-}F_i^{/}$. Between the distant peaks located on the opposite planes of sugar molecules, the location of flat nitrogenous bases is difficult, due to the obstacles encountered in the form of sugar molecule atoms and covalent chemical bonds. Therefore, out of the 12 possible coordinate planes, only five coordinate planes designated by the vertices of the polytope

$$\frac{F_i^{/}H_{(1)}}{H_{(1)}^{/}F_i}, \frac{F_i^{/}C_{(2)}}{C_{(2)}^{/}F_i}, \frac{F_i^{/}H_{(3)}}{H_{(3)}^{/}F_i}, \frac{F_i^{/}C_{(1)}}{C_{(1)}^{/}F_i}, \frac{F_i^{/}OH}{(OH)^{/}F_i}. \tag{9.3}$$

If among 12 pairs (Spirin, 2019) canonical (Watson, Crick, 1953a, b) base pairs for linked DNA molecules are distinguished, then there are only 4 of these pairs

$$A:U, G:C, C:G, U:A. \tag{9.4}$$

If RNA molecules are linked, then in (9.4), base U must be replaced with base T.

Considering that there are 4 canonical base pairs, and there are 5 planes in which each pair can be located, the total number of base pairs location options is 20, i.e., exactly as many different amino acids exist. Thus, each variant of the arrangement of base pairs corresponds to a certain amino acid connected in the ribosome to transport RNA. As is known (Spirin, 2019), the attachment of an amino acid to transport RNA carries out a specific enzyme that recognizes the amino acid and its corresponding transport RNA when they are combined. In this case, transport RNA is associated with messenger RNA, and between these RNAs at the junction point there again exists a polytopic of hereditary information. Thus, information on the state of messenger RNA is transmitted by transport RNA, on which the given amino acid joins. As we see in this process, there is no need to use the concept of a genetic code to describe it. The latter was introduced as an attempt to explain the difference in the number of canonical base pairs from the number of amino acids. However, this concept together with the concept of the codon does not explain anything. Taking into account the dimension of the polytope of genetic information allows us to determine the internal (hidden) order in the location of the bases. The situation is similar to the situation with the discovery of quasicrystals. X-ray diffraction patterns of some intermetallic compounds revealed a lack of translational symmetry in them (Shehtman, 1984). This caused a stormy reaction of the scientific community. Such intermetallic compounds were called quasicrystals. However, in works (Zhizhin, 2014, 2018) it was shown that translational symmetry immediately appears if we consider diffractograms as projections onto a two -dimensional plane of a structure of higher dimension (a hidden order of diffractograms is revealed).

Currently, there are no results of special experiments to determine a unique correspondence between the location of this pair of bases in the polytopic of hereditary information and a specific amino acid. However, one can qualitatively confirm the existence of five different arrangements of base pairs using experimental data on the correspondence of codons (nucleotide triplets) to amino acids (Spirin, 2019). If we assume that the average nucleotide in the codon plays the main influence in this correspondence, then five amino acids correspond to triplets with an average nucleotide *U*: *Phe*, *Leu*, *Ile*, *Met*, *Val*. Since each amino acid corresponds to one of the positions of nucleotide *U*, it turns out that these directions exist 5. A similar situation is repeated with triplets with an average nucleotide *C*. They correspond to 5 amino acids: *Ser*, *Pro*, *Thr*, *Ala*, *Cus*. For triplets with an average nucleotide *G* there correspond the amino acids: *Ser*, *Cys, Trp*, *Arg*, *Gly*. Triplets with an average nucleotide A correspond to 7 amino acids: *Tyr*, *His*, *Glu*, *Asn*, *Lys*, *Gln*, *Asp*. Some violations of the strict unambiguous equality of 5 amino acid numbers to each nucleotide can be explained by the influence of lateral nucleotides in triplets and noncanonical parasitic bases.

It should be noted that taking into account the highest dimension of the polytopic of hereditary information is also necessary in the analysis of pandemics in which viral RNA penetrates the cell and forces the ribosomes to produce proteins necessary for viruses to reproduce. The antiviral drugs currently under development are acting on one or another phase of the development of a viral infection (Kozlov, Stragunsky, 2002; De Clerey, 2004; Razonable, 2011; Crotty, et al., 2002; Caly, et al., 2020). Currently, clinical trials of antiviral drugs based on nucleic acid analogs that are structurally similar to natural RNA and DNA but have altered any of the main components of nucleic acids: phosphate backbone, pentose sugar, or one of the canonical nucleic bases, are currently undergoing clinical trials. In particular, analog nucleic bases give, among other things, various properties of base pairing and base folding. This is especially significant in light of the established high dimension of the interaction region of nucleic acids. For example, there are universal bases that can mate with all four canonical bases and thereby stop the replication of viral RNA. An example of analog RNA is a peptide nucleic acid, which has a high binding strength and does not contain a faceted backbone. Analog nucleic acids are also called xenon nucleic acids. They represent one of the main pillars of xenobiology, the construction of new life forms based on alternative biochemistry. In all these cases of noncanonical nucleic acids, it is of interest to study their geometry taking into account the higher dimensionality of polytopes and to analyze the effect of higher dimensionality on the features of their interaction.

9.4. The Law of Conservation of Incidents in Polytope of Hereditary Information

The monograph (Zhizhin, 2019 a) introduced the concept of the incidence coefficients of elements of lower dimension with respect to elements of the higher dimension and elements of higher dimension with respect to elements of the lower dimension. The first characterizes the number of elements of a certain higher dimension to which the given element of a lower dimension belongs. The second characterizes the number of elements of a given lower dimension that are included in a particular element of a higher dimension. Here we must remember that the vertices of geometric elements of various dimensions are atoms, molecules or functional groups. Therefore, the incidence of geometric elements to a friend means contact between particles of the matter, including living matter. The contact between particles of matter can be interpreted as the transfer of information on material structures, including biological structures. We introduce the notation: $k_{d^i_j d_u}$ the number of elements of dimension u, which include an element of dimension j ($u > j$) with number i. Thus, $k_{d^i_j d_u}$ is the incidence factor of element i with dimension j relative to elements with dimension u. We introduce the notation also: $k_{d_j d^i_u}$ j, which included in element i with a dimension u ($u > j$). Thus, $k_{d_j d^i_u}$ incidence factor of element i with dimension u relatively to elements with dimension j.

The smallest dimension of the cross-polytope is 4. From (9.2) it follows that in this polytope there are 8 vertices, 24 edges, 32 two-dimensional faces,

16 three-dimensional faces (Figure 9.2). The factors of incidents (from smaller dimension to larger) are

$$k_{d_0^i d_1} = 6, i = 1 \div 8; k_{d_0^i d_2} = 12, i = 1 \div 8; k_{d_0^i d_3} = 8, i = 1 \div 8; k_{d_0^i d_4} = 1, i = 1 \div 8; k_{d_1^i d_2} = 4, i = 1 \div 24;$$
$$k_{d_1^i d_3} = 4, i = 1 \div 24; k_{d_1^i d_4} = 1, i = 1 \div 24; k_{d_2^i d_3} = 2, i = 1 \div 32; k_{d_2^i d_4} = 1, i = 1 \div 32; k_{d_3^i d_4} = 1, i = 1 \div 16.$$

Sum up the incidence coefficients for all vertices, edges, two-dimension faces and three -dimension faces of the 4-cross-polytope

$$\sum_{i-1}^{8} k_{d_0^i d_1} + \sum_{i-1}^{8} k_{d_0^i d_2} + \sum_{i-1}^{8} k_{d_0^i d_3} + \sum_{i-1}^{8} k_{d_0^i d_4} + \sum_{i-1}^{24} k_{d_1^i d_2} + \sum_{i-1}^{24} k_{d_1^i d_3} + \sum_{i-1}^{24} k_{d_1^i d_4} +$$
$$\sum_{i=1}^{32} k_{d_2^i d_3} + \sum_{i=1}^{16} k_{d_3^i d_4} + \sum_{i=1}^{32} k_{d_2^i d_4} = 544. \tag{9.5}$$

The factors of incidents (from larger dimension to smaller, Figure 9.2) are

$$k_{d_0 d_1^i} = 2, i = 1 \div 24; k_{d_0 d_2^i} = 3, i = 1 \div 32; k_{d_0 d_3^i} = 4, i = 1 \div 16; k_{d_0 d_4} = 8; k_{d_1 d_2^i} = 3, i = 1 \div 32;$$
$$k_{d_1 d_3^i} = 6, i = 1 \div 16; k_{d_1 d_4} = 24; k_{d_2 d_3^i} = 4, \mathrm{i} = 1 \div 8; k_{d_2 d_4} = 32; k_{d_3 d_4} = 16.$$

Sum up the incidence coefficients for all elements of the 4-cross-polytope with dimension larger of zero

$$\sum_{i=1}^{24} k_{d_0 d_1^i} + \sum_{i=1}^{32} k_{d_0 d_2^i} + \sum_{i=1}^{32} k_{d_1 d_2^i} + \sum_{i=1}^{16} k_{d_0 d_3^i} + \sum_{i=1}^{16} k_{d_1 d_3^i} + \sum_{i=1}^{16} k_{d_2 d_3^i} + k_{d_0 d_4} + k_{d_1 d_4} + k_{d_2 d_4} + k_{d_3 d_4} = 544. \tag{9.6}$$

Comparing (9.5) and (9.6) you can see that the sum of incidents in a 4-cross-polytope from elements with a lower dimension to elements with a higher dimension is equal to the sum of incidents from elements with a higher dimension to elements with a lower dimension. Thus, the sum of incidents retains its value when changes the direction of the relationship between the elements (the law of conservation of incidents).

9.4.1. Theorem 9.1

In any cross-polytope of dimension n, the sum of all incidents of elements of a lower dimension with respect to all elements of a higher dimension is equal to the sum of all incidents of elements of a higher dimension with respect to all elements of a lower dimension and equals the sum of the series

$$f_0(n)3^{n-1} + f_1(n)3^{n-2} + f_2(n)3^{n-3} + ... + f_{n-1}(n)3^0, f_d(n) = C_n^{n-1-d} 2^{1+d}, d = 0 \div (n-1).$$

Proof

According to equation (9.2), each cross-polytope of dimension n has $2n = f_0(n)$ vertices. The peculiarity of a cross-polytope is that each of its vertices has an opposite vertex, with which it is not connected by an edge. Moreover, there is one edge between this vertex and all other vertices. We subtract from the total number of vertices two vertices (the selected vertex and its opposite) 2n-2. This is the possible number of edges emanating from the selected vertex. Thus, the incidence coefficient of the edges of any vertex is $k_{d_0^i d_1} = 2(n-1) = 2C_{n-1}^1, i = 1 \div f_0(n).$

be depicted as a projection on a two-dimensional plane (Zhizhin, 2019 a). In this image, all its vertices are located on a circle, with the selected vertex and its opposite vertex located symmetrically relative to the center of the circle. A mentally drawn line through these two vertices halves the circle (Figure 9.4). Therefore, the number of variants the location of the edges from the selected vertex to the other vertices in one of the halves of the circle is n-1, i.e., the number of combinations of n-1 vertices one by one. Since there are two halves of a circle, then for the total number of vertex selection options for edge formation, this number of combinations should be multiplied by 2. This is the meaning of the expression for the incidence rate of any vertex in the n-cross-polytope with respect to the edge. To further prove the theorem 1 and clarify the nature of the formation of the coefficients of incidence in the n-cross-polytope, we will use this technique. We arrange the vertices of the 4-cross-polytope in two lines, so that the vertices unlinked by an edge form vertical pair: each vertex in the top line is not connected by an edge to the vertex in the bottom line located strictly under this vertex in the top line

$$d_0^1\ d_0^2\ d_0^3\ d_0^4$$
$$d_0^5\ d_0^6\ d_0^7\ d_0^8 \qquad (9.7)$$

We choose some vertex in the top line (9.7), for example d_0^1

be connected by an edge with a vertex d_0^5

connected by an edge with other vertices of the bottom line. The number of such options is 3, i.e., with $n = 4$ is $3 = C_{n-1}^1 = C_3^1$ d_0^1 may be connected by an edge with any of the remaining three vertices in the top line (9.7). Therefore, the incidence coefficient of a vertex with respect to an edge in 4-cross-polytope $k_{d_0^i d_1} = 2C_{n-1}^1 = 6, i = 1 \div 8.$

belonging of a vertex d_0^1

determine the number of options for the participation of two vertices in the top line (9.7) (without a vertex d_0^1

vertex d_0^5 $2C_3^1$. In addition, as the vertices of the triangle, there may be vertices located in the upper and lower lines (9.7) in a cross-section way. There are 6 such variants in (9.7), i.e., more $2C_3^1$. Therefore, the incidence coefficient of a vertex with respect to a two-dimensional element in 4-cross-polytope is $k_{d_0^i d_2} = 2^2 C_{n-1}^2 = 12, i = 1 \div 8.$ Multiplying this number by the number of vertices in 4-cross-polytope you can get the total number of incidents of vertices to two-dimensional elements in 4-cross-polytope equal to 96. This number coincides with the corresponding number, defined earlier in Figure 9.4. Combination d_0^1 with three the different vertices from (9.7) can get the incidence coefficient of vertex to the three-dimension body (tetrahedron) for condition absent in combination opposite vertices. Can show that this combination is 8, $k_{d_0^1 d_3} = 2^3 C_3^3 = 8, i = 1 \div 8.$ Multiplying this number by the number of vertices in 4-cross-polytope you can get the total number of incidents of vertices to three-dimensional elements in 4-cross-polytope equal to 64. This number coincides with the corresponding number, defined earlier in Figure 9.4. Combination two vertices, for example d_0^1 and d_0^2 $d_0^1\ d_0^2$) with two the different vertices from (9.7) you can get the incidence coefficient of edge to three-dimension body (tetrahedron) for condition absent in combination opposite vertices. Can show that this combination is 4, $k_{d_1^1 d_3} = 2^2 C_2^2 = 4, i = 1 \div 8$. Multiplying this number by the number of edges in 4-cross-polytope you can get the total number of incidents of edges to three-dimensional elements in 4-cross-polytope equal to 96. This number coincides with the corresponding number, defined earlier in Figure 9.4. Combination two vertices (the edge $d_0^1\ d_0^2$

can get the incidence coefficient of edge to triangle for condition absent in combination opposite vertices. Can show that this combination is 4, $k_{d_1^1 d_2} = 2C_2^1 = 4, i = 1 \div 8$. Multiplying this number by the number of edges in 4-cross-polytope you can get the total number of incidents of edges to triangle in 4-cross-polytope equal to 24. This number coincides with the corresponding number, defined earlier in Figure 9.4. In general case can write of $2n$ vertices of the n-cross-polytope in two lines

$$\begin{matrix} d_0^1 & d_0^2 & d_0^3 & \dots & d_0^n \\ d_0^5 & d_0^6 & d_0^7 & \dots & d_0^{2n} \end{matrix}, \tag{9.8}$$

where vertex in the top line is not connected by an edge to the vertex in the bottom line located strictly under this vertex in the top line. The incidence coefficient of a vertex with respect to an edge in n-cross-polytope is $k_{d_0^i d_1} = 2\mathrm{C}_{n-1}^1, i = 1 \div f_0(n)$. The incidence coefficient of a vertex with respect to a two-dimensional element in n-cross-polytope is $k_{d_0^i d_2} = 2^2\,\mathrm{C}_{n-1}^2, i = 1 \div f_0(n)$. The incidence coefficient of a vertex with respect to a three-dimensional edge in n-cross-polytope is $k_{d_0^i d_3} = 2^3\,\mathrm{C}_{n-1}^3, i = 1 \div f_0(n)$. Go on can to say the incidence coefficient of a vertex with respect to a (n-1)-dimensional element in n-cross-polytope $k_{d_0^i d_{n-1}} = 2^{n-1}\,\mathrm{C}_{n-1}^{n-1}, i = 1 \div f_0(n)$. Obviously, that the incidence coefficient of a vertex with respect to n-cross-polytope is $k_{d_0^i d_n} = 1, i = 1 \div f_0(n)$.

Multiplying of the incidence coefficients of a vertex with respect to different dimension elements in n-cross-polytope and sum the product you can get the common express for the number of incident vertices to elements of different dimension in a n-cross-polytope

$$f_0(2^0+2^1C_{n-1}^1+2^2C_{n-1}^2+...+2^{n-1}C_{n-1}^{n-1})=f_0(n)\sum_{i=0}^{n-1}2^iC_{n-1}^i=f_0(n)3^{n-1}.$$

In this way you can get the common express for the number of incident edges to elements of different dimension more one in n-cross-polytope

$$f_1(2^0+2^1C_{n-2}^1+2^2C_{n-2}^2+...+2^{n-2}C_{n-2}^{n-2})=f_1(n)\sum_{i=0}^{n-2}2^iC_{n-2}^i=f_1(n)3^{n-2},$$

and go on.

In result, one can get the common express for the sum of all incidents of elements of a lower dimension with respect to all elements of a higher dimension in a n-cross-polytope

$$f_0(n)3^{n-1}+f_1(n)3^{n-2}+...+f_{n-1}3^0=\sum_{i=0}^{n-1}f_i(n)3^{n-1-i},\ f_i(n)=C_n^{n-1-i}2^{1+i}. \tag{9.9}$$

Let $k_{d_h d_j^i}$ be the number of elements of dimension h in a n-cross-polytope belonging to some one element of dimension j $(h<j)$, that is d_j^i. Obviously, this number is equal to $f_h(j)_s$ for simplex, and $i=1\div f_j(n)_{cr}$ for n-cross-polytope. So, elements of a cross-polytope are simplexes the product $f_h(j)_s f_j(n)_{cr}$

h belonging to all elements of dimension j for a simplex. This product is equal to $f_d(n)=C_{n+1}^{d+1}$

$$C_{j+1}^{h+1}2^{1+j}C_n^{n-1-j}=2^{1+j}\frac{(j+1)!}{(h+1)!(j-h)!}\frac{n!}{(n-1-j)!(1+j)!}.$$

Let us compare this number with the number of elements of dimension j, which have elements of dimension h in the n-cross-polytope

$$k_{d_h^i d_j}f_h(n)_{cr}=C_n^{n-1-h}2^{1+h}C_{n-h-1}^{j-h}2^{j-h}=2^{1+j}\frac{n!}{(n-1-h)!(1+h)!}\frac{(n-h-1)!}{(n-j-1)!(j-h)!}.$$

Obviously, these numbers are equal to each other. This proves that the number of elements of dimension h in a n-cross-polytope belonging to all dimensions of j $(j > h)$ n-cross-polytope is equal to the number of elements of dimension j, which have elements of dimension h as elements. Thus, the total number of incidents of elements of a smaller dimension with respect to elements of a higher dimension is equal to the total number of incidents of elements of a higher dimension with respect to elements of a smaller dimension. The total number define expression (9.9). Q.E.D.

The polytope of hereditary information is a cross-polytope of dimension 13. Substituting the value n = 13 and the values of $f_i(n)$ previous section in expression (9.9), we find the total value of the incident flow in the polytope of hereditary information

$$\sum_{i=0}^{12} f_i(13)3^{12-i} = 1.78 \cdot 10^9 \cdot$$

A significant value of the total incidence stream in the polytope of hereditary information indicates an intensive exchange of information between elements of the polytope of hereditary information. For example, this value is 400 times larger than the incident flux in a simplex of dimension 13 (Zhizhin, 2019 d). This may explain the recently discovered epigenetic principle of the transmission of hereditary information without changing the sequence of genes in DNA and RNA molecules.

9.5. Methylated Polytope of Hereditary Information

Recently, the question of counteracting viral infections, their spread and development has become acute. Various stages of the development viruses from capsids to return to capsids are studied (Agol, et al., 1990; Novikova, et al., 2002; Novikova, 2007; Spirin, 1990; Wurm, et al., 2001; Wulan, et al., 2015; Lundberg, et al., 2013; Crotty, et al., 2002). Various drugs are being developed to inhibit various stages of viral population development (Gonzalez, et al., 2008; Dong, et al., 2020; De Clerey, 2004; Crotty et al., 2002; Caly et al., 2012; Razonable, 2011). However, we are forced to state that viruses have amazing properties of attacking a living organism, penetrating it, suppressing the protective properties of the organism, using the life-supporting functions of the organism to its advantage, reproducing inside the organism, often leading ultimately to the death of the organism. The drugs developed against viruses can often only temporarily extinguish the infection, and viruses are modified and, in a new form, attack living organisms again. What allows viruses to often be the winner in the fight against living organisms? It should be remembered here that viruses are equal ancient players in the development of life. It is assumed that they preceded the emergence of universal LUCA (Last Universal Common Ancestor) protocells, since structural and functional similarities were found between archaeal viruses and some DNA-containing eukaryotic viruses (Novikova, 2007). This leads to the idea that it was viruses that contributed to the creation of life on Earth.

During their existence (about 5 billion years), viruses have developed complex systems of control, information transfer, recognition of dangers and ways to overcome them. Viruses participated in the creation of DNA and are themselves carriers of genetic information. Communication of information is always an important issue in a complex self-regulating system.

Following the model of Watson and Crick (Watson and Crick, 1953 a, b), the transfer of genetic information in a double-stranded DNA molecule occurs as a result of the complementary nitrogenous bases of the two helices of the molecule. At the same time, the genetic information is incorporated not only in the sequence of arrangement along the spirals of various nitrogenous bases (nucleotide), but also in the geometric characteristics of complementary compounds: distances between atoms in compounds (centers of glycosidic bonds, locations of glycosidic centers, angles between glycosidic bonds) (Spirin, 2019).

It was recently discovered (Zhizhin, 2019 a, b) that the vicinity of nitrogenous bases at the place of their connection in DNA is a polytope of dimension 13 (polytope of hereditary information), on the coordinate two-dimensional planes of which nitrogenous bases can be located. In this case, the geometric characteristic of the complementary compound of nitrogenous bases is the geometric characteristic of the structure of the hereditary information polytope. The main parameter here is incidence (Zhizhin, 2019 c), i.e., belonging of one element of some dimension of the polytope to other elements of another dimension. The structure of the polytope will include a listing of all such incidents. The law of conservation of incidents was established (Zhizhin, 2019 d), according to which the integral sum of incidents from elements of low dimension to elements of high dimension is equal to the integral sum of incidents from elements of high dimension to elements of low dimension. An analytical expression for calculating the integral sum of incidents over the polytope is obtained. The integral sum of incidents over the polytope is interpreted as the flow of information between the elements of the polytope. It is shown that an increase in the dimension of the polytope leads to a sharp increase in the value of the information flow between the elements of the polytope.

A distinctive feature of the DNA of a number of viruses is the presence of methylated nitrogenous bases in them, i.e., a methyl group is attached to nitrogenous bases. For example, in bacterial DNA methylated bases are 5′-methylcytosine, 6′-methylaminopurine (Novikova, 2007; Agol, et al., 1990). The origin of such bases is the enzymatic methylation of the already synthesized DNA strand. This process is carried out by virus-specific methylases, which use the methyl groups of S-adenosylmethionine of the host cell as a donor (Novikova, 2007).

In this section, the effect of methylation of nitrogenous bases on an increase in the intensity of transmission of hereditary information is theoretically investigated. This is due to an increase in the dimension of the hereditary information polytope during its methylation. It was shown that the flow of information during methylation of the hereditary information polytope increases 2000 times as compared with the flow of information in the absence of methylation. This may explain the high degree of organization of viruses when they attack living organisms.

It should be noted that the experimentally intensifying effect of methylation on the transmission of hereditary information is currently observed in plants (Hawkes, et al., 2016; Lindquist, et al., 2016; Mancuso, 2017).

9.6. Nucliec Acids Methylation

The concept of nucleic acids as chains containing only four types of nucleotides is somewhat simplified. In DNA, as follows from various sources, there is a certain number of methylated nitrogen bases, i.e., bases containing a methyl group-CH_3. For example, the Fisher formula of cytosine $C_4H_5N_3O$ has the form (Figure 9.5).

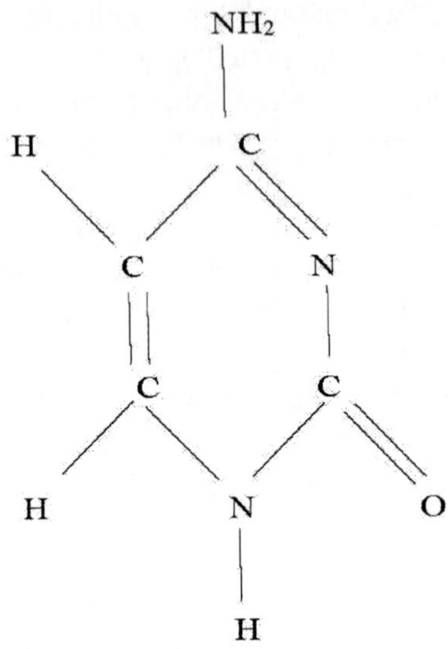

Figure 9.5. Fisher's formula for cytosine.

Fisher's formula for methylated cytosine (5-methylcytosine) is shown in Figure 9.6.

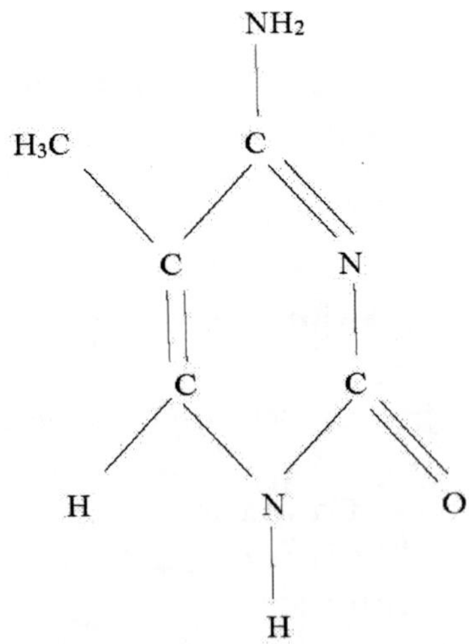

Figure 9.6. Fisher Formula for Methylated Cytosine.

The nitrogenous base adenine (aminopurin) $C_5H_5N_5$ has the Fisher formula (Figure 9.7).

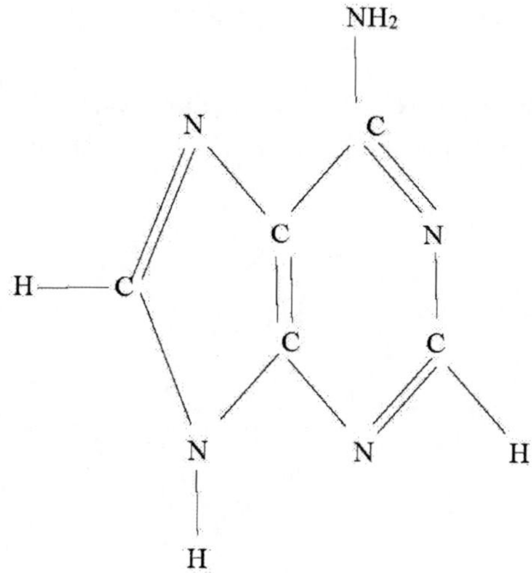

Figure 9.7. Fisher Adenine Formula Methylated adenine (6-methyladenine) has the Fisher formula (Figure 9.8).

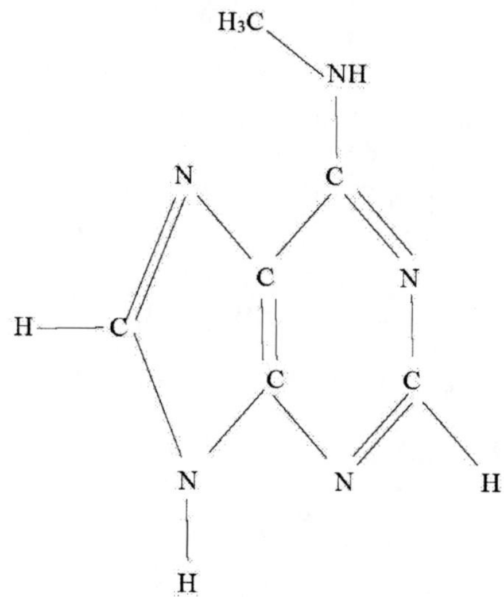

Figure 9.8. Fisher Formula for Methylated Adenine.

Methylation occurs after polynucleotide synthesis. It is used by organisms in its fight against viruses. However, viruses also use methylation to counteract the defenses of organisms. For example, viruses in nucleic acids contain oxymethylcytosine instead of cytosine (Figure 9.9).

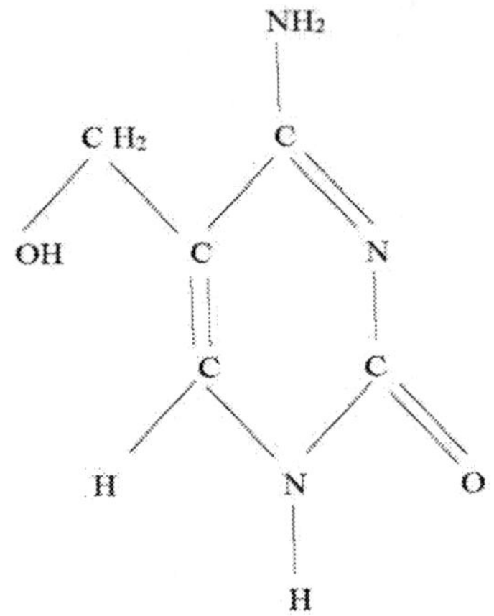

Figure 9.9. Fisher's Formula of Oxymethylcytosine.

Here, a hydroxyl group is included in the methyl groups. Emerging hydroxyl groups may carry one or two glucose residues.

One of the bacteriophages that infect *Bacillus subtitis* contains 5-hydroxymethyluracil instead of uracil, and 5-dioxipentyluracil instead of thymine (Figure 9.10).

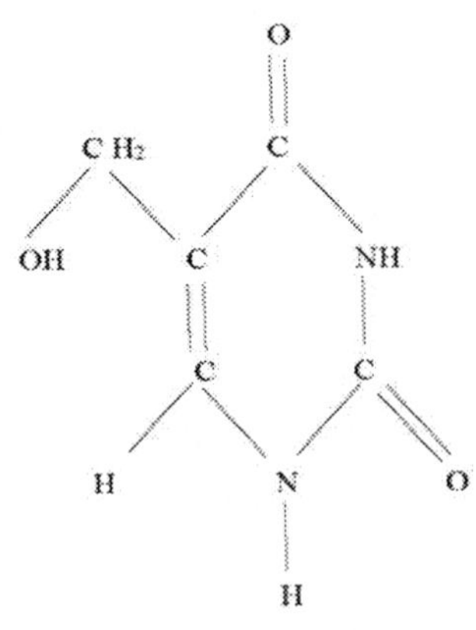

Figure 9.10. Fisher's Formula 5-Dioxipentylluracil.

Instead of guanine (Figure 9.11) as a result of methylation, O-methylguanine is formed (Figure 9.12).

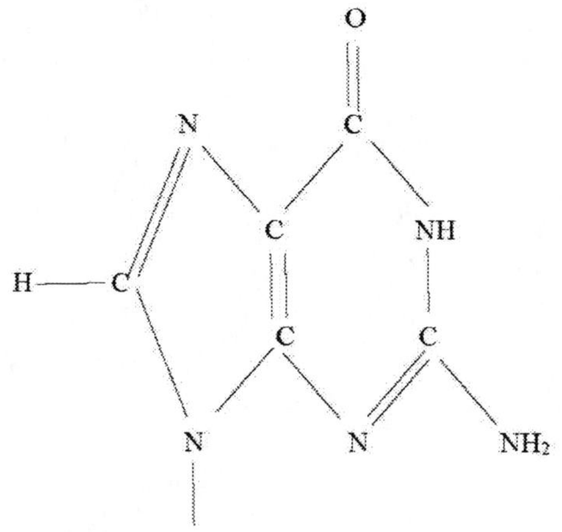

Figure 9.11. Fisher`s Formula Guanine.

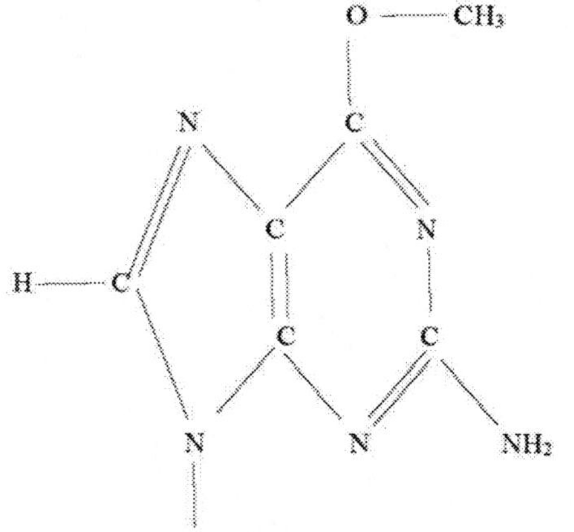

Figure 9.12. Fisher's Formula O^6-Methylguanine.

Thus, all classic nitrogenous bases as a result of methylation can take a different form. Moreover, it was established that these changes are very significant for the opposing cells of organisms and viruses.

In eukaryotic DNA, about 5% of the total cytosine content is $5^{/}$-methylcytosine. The degree of DNA methylation changes during development, and also depends on the age of the body and hormonal influences. DNA in vertebrates is especially strongly methylated. Methylation can accompany DNA replication; it is inherited and transmitted to offspring. Modified DNA can affect the local structure of the chromosome. Methylation promotes the transition to DNA Z-conformation, controls the activity of genes (Spirin, 1990). However, the mechanism of action of methylation has not yet been disclosed.

Thus, all nucleotides of the classical type can, under appropriate conditions, be replaced by nature itself with modified nucleotides with methylated nitrogen bases. In this case, the complementarity of the base compound is preserved, and the symmetry of nucleic acids is preserved. What is the secret to the need for such modifications of nitrogen bases? Try to answer this question. It was proved that many biomolecules have a higher dimension (Zhizhin, 2016, 2018, 2019 e). In particular, it was proved that the connection of a carbon atom with four other atoms or functional groups has a dimension of 4. A methyl group attached to nitrogen bases is a special case of such a compound. Consequently, methylation changes the dimension of nitrogenous bases. Without a methyl group, the dimension of nitrogen bases is 2, with the addition of a methyl group, the dimension of nitrogen bases becomes 4 or more. Find out what this leads to.

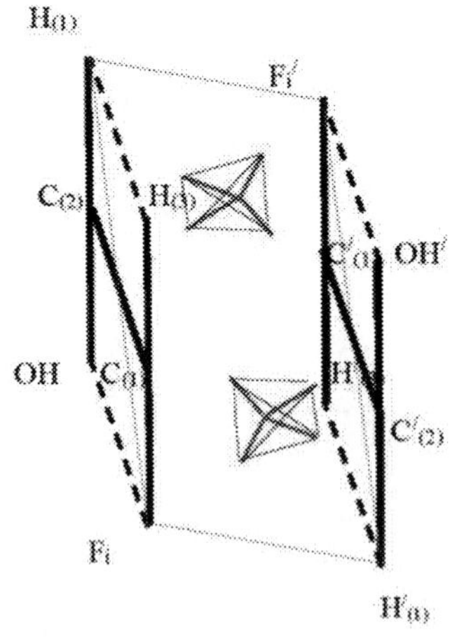

Figure 9.13. The formation of four-dimensional polytopes in nitrogen bases during nucleation of nucleic acids.

When nucleic acids are labeled, their nitrogenous bases cease to be flat, since a tetrahedron with a center is formed in the vicinity of one of the carbon atoms of these bases, i.e., four-dimensional polytope. In this case, due to the antiparallel nature of molecular structures formed by antiparallel nucleic acid helices, it should be assumed that such a four-dimensional polytope appears in both linked nitrogen bases (Figure 9.13).

In Figure 9.18, the coordinate plane $\frac{F_i^{/} H_{(1)}}{H_{(1)}^{/} F_i}$ is selected as an example. In Figure 9.18, the edges of the tetrahedra with the center, marked in red, correspond to chemical covalent bonds. Thin black ribs are needed to create a closed convex body. The general construction of the polytope of hereditary information synchronizes the type of cross-polytope, but the number of vertices in it is 10 more than that of the non -methylated polytope of hereditary information. The complementarity of the nitrogenous base compound under these conditions is preserved, the symmetry of nucleic acids is preserved. This is facilitated by the mechanisms laid down by nature in nucleic acids that support complementarity and break the chains of nucleic acids when such conditions are violated (Spirin, 1990). According to (Zhizhin, 2019 c), there is a relation between f_i elements i in the cross-polytope and the dimension d of the cross-polytope itself

$$f_i(d) = 2^{1+i} C_d^{d-1-i}. \tag{9.10}$$

Using formula (9.10) from the number of vertices at which the dimension $i = 0$, we can calculate the dimension of the methylated polytope of hereditary information

$$f_0 = 2C_d^{d-1} = 2d = (26+10) = 36, d = f_0 / 2 = 18. \tag{9.11}$$

Thus, the methylated polytope of the hereditary information has dimension 18.

Consequently, the dimension of the methylated polytope of hereditary information increased by 5 units compared with the dimension of the unmethylated polytope of hereditary information (13). By the ratio (9.10), it is possible to determine the number of elements of different dimensions and in the methylated polytope of hereditary information.

The number of edges is $f_1(18) = 2^2 C_{18}^{16} = 612$

is $f_2(18) = 2^3 C_{18}^{15} = 6528$, the number of tetrahedrons is $f_3(18) = 2^4 C_{18}^{14} = 48960$, the number of fourth-dimensional faces (simplexes) is $f_4(18) = 2^5 C_{18}^{13} = 274176$

faces (simplexes) is $f_5(18) = 2^6 C_{18}^{12} = 1188096$

dimensional faces (simplexes) is $f_6(18) = 2^7 C_{18}^{11} = 4073472$

of seven-dimensional faces (simplexes) is $f_7(18) = 2^8 C_{18}^{10} = 11202048$

the number of eight-dimensional faces (simplexes) is $f_8(18) = 2^9 C_{18}^{9} = 24893440$, the number of nine-dimensional faces (simplexes) is $f_9(18) = 2^{10} C_{18}^{10} = 44808192$

dimensional faces (simplexes) is $f_{10}(18) = 2^{11} C_{18}^{11} = 65175552$

number of eleven-dimensional faces (simplexes) is

$f_{11}(18) = 2^{12} C_{18}^{12} = 76038144$, the number of twelve-dimensional faces (simplexes) is $f_{12}(18) = 2^{13} C_{18}^{13} = 70189056$,

thirteenth-dimensional faces (simplexes) is

$f_{13}(18) = 2^{14} C_{18}^{14} = 50135040$, the number of fourteenth-dimensional faces (simplexes) is

$f_{14}(18) = 2^{15} C_{18}^{15} = 26738688$, the number of fifteenth-dimensional faces (simplexes) is

$f_{15}(18) = 2^{16} C_{18}^{16} = 10027008$, the number of sixteenth-dimensional faces (simplexes) is

$f_{16}(18) = 2^{17} C_{18}^{17} = 2359296$, the number of eighteenth-dimensional faces (simplexes) is

$f_{17}(18) = 2^{18} C_{18}^{18} = 262144$.

The obtained numbers determine the structure of the methylated polytope of hereditary information (Zhizhin, 2021 b). To prove the existence of a methylated polytope of hereditary information, we use the Euler-Poincaré equation (9.1) (Poincaré, 1895)

Substitute the obtained values f_i

36-612 + 6528-48960 + 274176-1188096 + 4073472-11202048 + 24893440-44808192 + 65175552-76038144 + 70189056-50135040 + 26738688-10027008 + 2359296-262144 = 0.

Equality to zero of the familiar variables proves that the methylated polytope of hereditary information exists, its dimension is even and equal to 18.

9.7. The Law of Conservation of Incidents in the Methylated Polytope of Hereditary Information

The monograph (Zhizhin, 2019 c) introduced the concept of the incidence coefficients of elements of lower dimension with respect to elements of the higher dimension and elements of higher dimension with respect to elements of the lower dimension. The first characterizes the number of elements of a certain higher dimension to which the given element of a lower dimension belongs. The second characterizes the number of elements of a given lower dimension that are included in a particular element of a higher dimension. Here we must remember that the vertices of geometric elements of various dimensions are atoms, molecules or functional groups. Therefore, the incidence of geometric elements to a friend means contacted between particles of the matter, including living matter. The contact between particles of matter can be interpreted as the transfer of information on material structures, including biological structures. We introduce the notation: $k_{e_j^n e_u}$ the number of elements of dimension *u,* which include an element of dimension *j* (*u* > *j*) with number *n*. Thus, $k_{e_j^n e_u}$ element *n* with dimension *j* relative to elements with dimension *u*. We introduce the notation also: $k_{e_j e_u^n}$ *j*, which included in element *n* with a dimension *u* (*u* > *j*). Thus, $k_{e_j e_u^n}$ incidence factor of element *n* with dimension *u* relatively to elements with dimension *j*.

In work (Zhizhin, 2019 c) was proved state Theorem 9.1. According to this theorem in any cross-polytope of dimension *d*, the sum of all incidents of elements of a lower dimension with respect to all elements of a higher dimension is equal to the sum of all incidents of elements of a higher dimension with respect to all elements of a lower dimension and equals the sum of the series:

$$f_0(d)3^{d-1} + f_1(d)3^{d-2} + f_2(d)3^{d-3} + ... + f_{d-1}(d)3^0 = \sum_{i=0}^{d-1} f_i(d)3^{d-1-i},$$

$$f_i(d) = C_d^{d-1-i}2^{1+i}, i = 0 \div (d-1). \tag{9.12}$$

Substituting in the formula (9.12) the values of $f_i(d)$ $d = 18$ we find that the sum of the incidents (information flow) for the methylated polytope of hereditary information takes the value

$$\sum_{i=0}^{d-1} f_i(d)3^{d-1-i} = 3.8143 \cdot 10^{12}. \tag{9.13}$$

Recall that the value of the information flow for the unmethylated polytope of hereditary information (see Zhizhin, 2019 b) was $1.78 \cdot 10^9$. Thus, the methylation of nucleic acids led to an increase in the flow of information between elements of the polytope of hereditary information by more than 2000 times.

Conclusion

The geometry of the neighborhood of the compound of two nucleic acid helices with nitrogen bases has been investigated in detail. It is proved that this neighborhood is a polytope with anti -parallel edges of dimension 13 (13-cross-polytope). This polytope is called of the polytope of hereditary information. The geometry of the polytope of hereditary information is investigated. It is shown that in the flat coordinate planes of the polytope of hereditary information, there are flat complementary hydrogen compounds of the nitrogenous bases of two nucleic acid helices. It turned out that the number of these coordinate planes (12) is exactly as many as there are various options for hydrogen compounds of nitrogenous bases. Thus, in each of these coordinate planes one of the possible types of bonding of nitrogenous bases is located. Thus, the possible orientation of flat nitrogenous bases in the space of higher dimension is determined. An image of the polytope of hereditary information with a specific indication of its coordinate planes is constructed. The incidence of low-dimensional elements to the highest-dimensional elements of this polytope is studied, as well as the incidence of higher-dimensional elements to low-dimensional elements of this polytope. The value of the total incidence flows from low -dimensional elements to higher-dimensional elements and vice versa are determined. These values turned out to be equal (the law of conservation of incidents) and exceeding one billion. This destroys the idea of the correspondence of triplets to certain amino acids, introduced logistically, in order to avoid the contradiction between a small number of nucleotides and a large number of amino acids (Zhizhin, 2020 a, b). It turned out that the polytope of hereditary information has an internal (hidden) degree of freedom in the form of a different arrangement of pairs of flat nitrogen base in the space of higher dimension. Consideration of this internal degree of freedom is necessary when analyzing the transmission of hereditary information and analyzing the spread of viral infectious diseases.

Currently, clinical trials of antiviral drugs based on nucleic acid analogs that are structurally similar to natural RNA and DNA but have altered any of the main components of nucleic acids.

Analog nucleic acids represent one of the main pillars of xenobiology, the construction of new life forms based on alternative biochemistry. In all these cases of noncanonical nucleic acids, it is of interest to study their geometry taking into account the higher dimensionality of polytopes and to analyze the effect of higher dimensionality on the features of their interaction.

The incidence of low-dimensional elements to the highest-dimensional elements of this polytope is studied, as well as the incidence of higher-dimensional elements to low-dimensional elements of this polytope. The values of the total incidence flow from low-dimensional elements to higher-dimensional elements and vice versa are determined. These values turned out to be equal (the law of conservation of incidents) and exceeding one billion.

This indicates an intensive flow of information between the elements of the polytopic of hereditary information, ensuring the transmission of hereditary information. Which can serve, in particular, to explain the existence of the transmission of hereditary changes without changing the sequence of genes (epigenetics).

Chapter 10

Dimension of Substances and Life

Abstract

It is proved that water molecules, as the basis of living organisms and the source of the origin of life, have the highest dimension. Various types of biomolecules with the highest dimension are listed. The possibility of memory of water about the presence of toxic substances in it is analyzed analytically.

Keywords: water molecules, biomolecules, dimension, memory of water, atomic orbital

Introduction

Life began in water and continues in water. Water is matter and the matrix of life. Water is not only outside of us, but also inside us. Every living organism is 80% water. Recent studies (Qian, et al., 2010; Russel, et al., 2005; Simakov, et al., 2020) show that as a result of the interaction of water with pyrite FS_2, there is a possibility of the appearance of protocells LUCA in water layers on the surface of minerals, including on the surface of nanodiamonds, which can form under terrestrial conditions at moderate temperatures and pressures (Simakov, 2018). All biomolecules ultimately evolved from a protocell that emerged over 4.5 billion years ago. The author's works (Zhizhin, 2019 a, b, c; 2018; 2020 a, b, c) prove that various biomolecules in nature have the highest dimension. This article proves that the water molecules themselves, from which biomolecules originated in contact with minerals, have the highest dimension. Thus, the highest dimension of biomolecules is a consequence of the highest dimension of water molecules. Water has the ability to adjust its structure to the structure of substances that are in it or border it. As an illustration of this property, the process of restructuring of water to match the structure of cyanide compounds dissolved in water is considered.

10.1. The Structure of Water

The water molecule has a peculiar shape. Oxygen in the second energy level *p* orbital has one quantum cell with a pair of electrons with opposite spins and two quantum cells with one electron in each of them. These two electrons combine with two electrons of the hydrogen atoms to form a covalent chemical bond. In addition, the oxygen atom has one more pair of electrons at the second energy level in the *s* orbital. The hydrogen atoms, which donated their two electrons to form a bond, remain sufficiently distant from the oxygen atom and have a positive charge +. Unshared electron pairs of the oxygen atom of the outer energy level naturally have a negative charge -. Conventionally, the shape of a water molecule can be represented as shown in Figure 10.1.

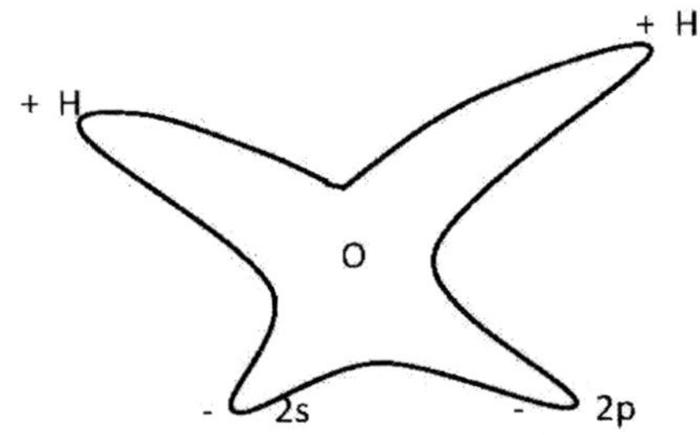

Figure 10.1. The shape of a water molecule.

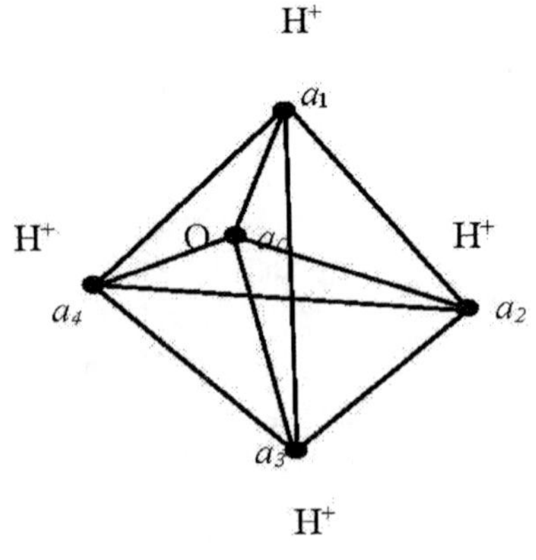

Figure 10.2. The tetrahedral structure with a center of molecules water with joined two protons.

It is obvious that positively charged protons of other water molecules can join the orbitals with negative charges of the second energy level of the oxygen atom. The result is a tetrahedral structure with a center (Figure 10.2).

The tetrahedron with a center on Figure 10.2 has 5 vertices $a_0 \div a_4$, 10 edges $a_0a_1, a_0a_2, a_0a_3, a_0a_4, a_1a_2$, $a_3a_1, a_1a_4, a_4a_2, a_2a_3, a_4a_3$, dimensional faces $a_3a_4a_1, a_1a_3a_2, a_1a_4a_2, a_2a_3a_4, a_1a_0a_3, a_1a_0a_2$, $a_0a_4a_1, a_0a_4a_2, a_0a_3a_2, a_0a_3a_4$, 5 three-dimensional faces $a_3a_4a_1a_2, a_0a_1a_3a_2, a_1a_4a_2a_0, a_3a_4a_1a_0, a_3a_4a_0a_2$.

To determine the dimension of a tetrahedron with a center, it is sufficient to use the Euler-Poincaré equation (Poincaré, 1895)

$$\sum_{i=0}^{n-1}(-1)^i f_i(P) = 1 + (-1)^{n-1}. \tag{10.1}$$

There n is dimension of a polytope P, f_i is the number of elements with dimension i in the polytope P.

In this case $f_0 = 5, f_1 = 10, f_2 = 10, f_3 = 5$.

Substituting the obtained values into equation (10.1), we obtain, i.e., Euler-Poincaré's equation for a tetrahedron with a center holds for n = 4. This proves that a tetrahedron with a center has dimension 4.

The existence of a tetrahedral distribution of hydrogen atoms in the network of hydrogen bonds of water molecules is proved by the methods of computer experiment (Eisenberg, Kautsman, 1975; Lyashchenko, Dunyashev, 2003; Rahman, Stillinger, 1973). However, the possibility of the existence of non-tetrahedral dimers is noted. This is quite natural since hydrogen bonds do not differ in strength and chaotic rearrangements of the network of hydrogen bonds are possible in water. However, the tetrahedral distribution of hydrogen atoms in the structure of water is predominant. It should be remembered that oxygen atoms are in the center of the tetrahedron, and the tetrahedron with the center has dimension 4 (this is not mentioned in all works on the structure of water). Note that this is significant if we remember the statement that life originated in water.

10.2. The Dimensional of Biomolecules

In biomolecules, you can find many examples of compounds in the form of a tetrahedron with a center, that is, similar to the structure of a water molecule (for example CH_4 , NH_4, PO_4 , etc.). All of them have a dimension of 4. The most important representatives of biomolecules are molecules of carbohydrates, proteins, lipids, nucleic acids. When depicting biomolecules, schematic planar images of Fisher and Haworth (Metzler, 1980; Lehninger, 1982; Koolman, Roehm, 2013) are usually used. But they do not reflect the real spatial structure of biomolecules. In 2016, the author proposed spatial images of biomolecules in the form of polytopes (Zhizhin, 2016), the dimension of which is determined by the Euler-Poincaré equation (10.1). Since the concept of a functional group is widely used in organic compounds, when determining the structure of biomolecules and their dimensions, not only individual atoms, but also functional groups can be located at the vertices of the corresponding polytope. This is how the concept of the functional dimension of a biomolecule was introduced. As a result, the functional dimensions of many biomolecules were determined. In particular (Zhizhin, 2018, 2019 a), the functional dimension of the simplest carbohydrate monosaccharide aldose (with three carbon atoms) turned out to be 5, as well as the dimension of the tartaric acid molecule. The functional dimension of the D-ribose molecule is 12. The functional dimension of the α – D-glucose molecule is 15. The functional dimension of any amino acid is 4, as is the dimension of phosphoric acid. Biomolecules connecting in series with each other form chains: polysaccharides, polypeptides, nucleic acids. Geometrically, they represent chains of polytopes of the highest dimension. Additional bonds between sections of the chain or between different polymer chains with the help of hydrogen bonds or metal ions (iron, calcium, magnesium, zinc) lead to the formation of polytopes of even greater dimension. Such conformations of biomolecules play a very important role in living organisms, organizing the processes necessary for the vital activity of organisms. It has just been established (Zhizhin, 2019 a, b, c; 2020 a, b) that the bond of nucleic acids by means of a hydrogen bond of nitrogenous bases leads to the formation of a polytope of hereditary information of dimension 13, characterized by a powerful flow of information exchange between the elements of this polytope. This makes it possible to explain the existence of the so-called epigenetic process of the transfer of hereditary information, which is not associated with a change in the sequence of genes.

10.3. The Memory of Water

There is no doubt about the existence of a certain relaxation time required for the water to return to its original state after the cessation of external influences, for example, a magnetic or electric field (Klassen, 1973; Martynova, et al., 1969; Tatarsky, 1989; Lalalyants, 1990). Let us try to explain the existence of the memory of water about the chemical compounds dissolved in it, which are essential for the life of biological organisms. Let such compounds be cyanide compounds. It is well known that hydrocyanic acid *HCN* and its salts are strong poisons. The toxic effect of these compounds is that the ions CN^- formed in their solution inhibit the reduction of oxygen by the most important respiratory enzyme, cytochrodoxidation, in the tissue cells of living organisms (Lehninger, 1982; Ershov, 1989). Ions CN^- react with the oxidized form of cytochrome oxidase and form a complex compound with a trivalent iron atom of enzyme $\left[Fe(CN)_6\right]^{3-}$. As a result of the overlapping of the orbitals of the complexing agent Fe^{3+} by the ligands CN^- and the hybridization of the orbitals, the ion $\left[Fe(CN)_6\right]^{3-}$ has octahedral coordination (Figure 10.3). Geometrically, this ion is an octahedron with a center.

The functional dimension of this ion is 4. The octahedron with a center on Figure 10.3 has 7 vertices $a_0 \div a_6, (f_0 = 7)$; 18 edges $a_0a_1, a_0a_2, a_0a_3, a_0a_4, a_0a_5, a_0a_6, a_1a_2, a_2a_3, a_3a_4, a_1a_4,$ $a_6a_1, a_6a_2, a_6a_3, a_6a_4, a_1a_5, a_5a_2, a_5a_3, a_4a_5, (f_1 = 18)$; 20 two-dimensional faces $a_0a_6a_1, a_0a_6a_2, a_0a_3a_6, a_0a_6a_4, a_1a_0a_5, a_5a_0a_2,$ $a_3a_0a_5, a_0a_4a_5, a_1a_0a_2, a_2a_3a_0, a_4a_0a_3, a_1a_0a_4,$ $a_5a_2a_1, a_3a_5a_2, a_5a_3a_4, a_5a_1a_4,$ $a_6a_2a_1, a_6a_3a_2, a_6a_3a_4, a_6a_1a_4, (f_2 = 20)$; 9 three-dimensional faces $a_0a_4a_1a_5, a_0a_1a_5a_2, a_3a_5a_2a_0, a_3a_4a_5a_0, a_1a_4a_0a_6, a_0a_1a_2a_6,$ $a_3a_6a_0a_2, a_0a_4a_3a_6, a_1a_4a_2a_5a_3a_6, (f_3 = 9)$.

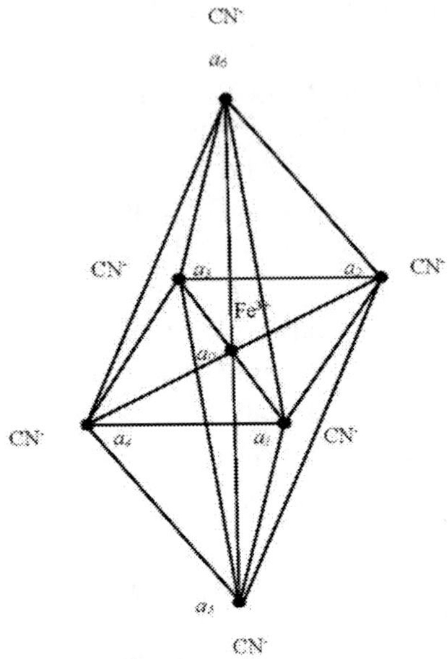

Figure 10.3. Geometric structure of the ion $[Fe(CN)_6]^{3-}$.

Substituting the obtained numbers $f_i (i = 0 \div 3)$

obtain,

$$7 - 18 + 20 - 9 = 0,$$

i.e., the Euler-Poincaré equation holds for n = 4. This proves that the octahedron is centered, that is, ion $[Fe(CN)_6]^{3-}$ has dimension 4. In this compound, each pair of electrons 2s of the nitrogen atom in each of the six ions CN^- occupies one of the six vacant quantum cells of the ion Fe^{3+}. The ion Fe^{3+} has two vacant quantum cells in orbit 3d, one vacant quantum cell in orbit 4s and three vacant quantum cells in orbit 4p. Hybridization of the orbitals sp^3d^2

that anions CN^- Fe^{3+} in comparison with other ions or molecules in the stoichiometric series in terms of the strength of interaction with the complexing agent. The strong bond of ions CN^- with cations Fe^{3+} disrupts the redox processes in cytochrome oxidase, consisting

in the transfer of electrons to copper atoms and then to oxygen. As a result, the ions H^+ formed in the body during the decomposition of proteins, carbohydrates and amino acids do not react with oxygen to form water. The concentration of ions H^+ in the body increases sharply, which leads to poisoning of the body. Obviously, in this way, the degree of poisoning will be determined by the concentration of ions CN^-
formed during the dissociation of hydrocyanic acid or its salts. Moreover, hydrocyanic acid dissociates weakly, and the dissociation of salts goes almost to the end. The dissociation of hydrocyanic acid as a result of the formation of ions H^+ slows down the dissociation of water. An increase in the concentration of hydrocyanic acid leads to a decrease in the degree of dissociation of both hydrocyanic acid and water. Indeed, the concentration of ions and molecules in solution is determined by the relations

$$\left[H^+\right]=c_{H_2O}\alpha_{H_2O}+c_{HCN}\alpha_{HCN},\left[OH^-\right]=c_{H_2O}\alpha_{H_2O},\left[CN^-\right]=c_{HCN}\alpha_{HCN},$$
$$\left[H_2O\right]=c_{H_2O}(1-\alpha_{H_2O}),\left[HCN\right]=c_{HCN}(1-\alpha_{HCN}). \tag{10.2}$$

There are α_{H_2O} – degree of dissociation of water, α_{HCN} – degree of dissociation of hydrocyanic acid, c_{H_2O} – concentration of water before dissociation, c_{HCN} – concentration of hydrocyanic acid before dissociation. Substituting relations (10.2) into the equations for the dissociation constants

$$K_{H_2O}=\frac{\left[H^+\right]\left[OH^-\right]}{\left[H_2O\right]},K_{HCN}=\frac{\left[H^+\right]\left[CN^-\right]}{\left[HCN\right]}, \tag{10.3}$$

taking into account that the degrees of dissociation are small compared to unity, we obtain

$$K_{H_2O}=c_{H_2O}\alpha_{H_2O}^2+c_{HCN}\alpha_{HCN}\alpha_{H_2O},$$
$$K_{HCN}=c_{H_2O}\alpha_{H_2O}\alpha_{HCN}+c_{HCN}\alpha_{HCN}^2. \tag{10.4}$$

Solving system (10.4) with respect to α_{HCN}, α_{H_2O}, taking into account that the mass fractions of water x_{H_2O} and acid x_{HCN}

$x_{H_2O} + x_{HCN} = 1, x_{HCN} << 1$

$$\alpha_{H_2O} = \frac{K_{H_2O}}{\sqrt{\rho(K_{H_2O} / \mu_{H_2O} + x_{HCN} K_{HCN} / \mu_{HCN})}}, \frac{\alpha_{HCN}}{\alpha_{H_2O}} = \frac{K_{HCN}}{K_{H_2O}}. \quad (10.5)$$

There are $\rho-$ solution density, $\mu-$ molecular weight.

Relations (10.2), (10.5) make it possible to calculate the degree of dissociation and ion concentration for a given initial concentration of hydrocyanic acid. In particular, taking the values of the dissociation constants $K_{H_2O} = 10^{-14}$ mol/l, $K_{HCN} = 4.79 \cdot 10^{-10}$ mol/l at very low concentrations of hydrocyanic acid $(x_{HCN} \to 0)$, we find $\alpha_{H_2O} = 10^{-8}, \alpha_{HCN} = 4.79 \cdot 10^{-4}$. At a lethal concentration of hydrocyanic acid (Ershov, 1989) $x_{HCN} 1.4 \cdot 10^{-6}$ these values are equal $\alpha_{H_2O} = 10^{-10}, \alpha_{HCN} = 4.6 \cdot 10^{-4}$. Thus, the dissociation of water in solutions of hydrocyanic acid and its salts is insignificant, and the degree of dissociation of hydrocyanic acid weakly depends on its concentration. From (10.2), (10.5) it follows that despite a decrease in the degree of dissociation of hydrocyanic acid with an increase in its concentration, the concentration of ions CN^- increases proportionally x_{HCN}

When any substances are dissolved, the solvent molecules are strongly influenced by the molecules of the dissolved substance. In this case, the solvent molecules are modified. It was shown (Semenchenko, 1932) that the water molecules surrounding the ions of the dissolved substance are under the influence of forces equivalent to a pressure of 50,000 atmospheres. In the process of hydration, molecules of hydrated water (Zundel, 1972) pass into an excited state with vibrations of nuclei relative to new acquired positions that differ from unexcited positions. The nature of the fluctuations is determined by the nature of the hydrated ion. This is how the nature of the hydrated ion is transferred to the water molecules in its environment. Due to the excited vibrations, water molecules become more deformable and polarizable. It is these properties that put the ion CN^- on the first place in the spectrochemical series of the activities of ions and the forces of their interaction with the complexing agent. The greater the polarizability of the particle, the stronger its bond with the complexing agent. The greater the deformability of the molecule, the easier it is for the molecule to "fit" into the particular relief of the enzyme (in this case, cytochrome oxidase). The network of hydrogen bonds in water has the ability to transfer perturbations external to the network to the entire volume of the solution, while maintaining the stability of the structure of the H – network (Rodnikova, 1993). The phenomenon of negative solvation is associated with this property (Samoilov, 1957), i.e., an increase in the mobility of solvent molecules near ions. All this indicates that the excitation created by ions CN^- with an increase in concentration is transferred to almost all water molecules in solution. After dilution of this solution, the acquired properties of water are retained for some time, which is necessary for the relaxation of the H-bond network. The characteristic time for a change in the properties of water in the absence of external influences on it is, according to experiments (Miller, 1969), several tens of minutes. While the characteristic time for the absorption of ions and water molecules into the tissue cells of living organisms is fractions of a second (Ivanov, 1993). It follows from this that water molecules, having changed under the influence of ions CN^- after dilution of water, can retain their activity and, instead of ions CN^- take their place as ligands around the complexing agent Fe^{3+} in the cytochrome oxidase enzyme, firmly binding with it. In this case, a pair of electrons 2s of the oxygen atoms in the water molecule is transferred to the vacant orbitals 3d, 4s, 4p of the iron atom of the complexing agent. Here, too, an octahedron with a center of dimension 4 is formed around the iron atom, and the violation of redox processes in cytochrome oxidase is preserved.

10.4. Chelated Compounds

One of the main factors determining the affinity of molecules for metal ions is the chelate effect, i.e., a clearly expressed ability of molecules to bind metal ions in the presence of groups capable of complex formation in the molecules. Nature has successfully used the chelate effect to create such important metal-containing molecules as porphins, chlorophyll, calcium-binding proteins, ribosomes, and amino acids (Metzler, 1980). During the formation of life on planet Earth, metal ions were formed during the interaction of metals in the earth's crust with water, leading to their spread over the surface of the planet (Fersman, 1923; Vernadsky, 1925). Most metal ions form a hydrate shell. These include not only iron ions, but also manganese, cobalt, nickel, zinc, and magnesium. These are all elements that play an important role in living organisms. As a rule, there are six water molecules in the hydration shell of metal ions and an octahedral structure with a center is formed. The dimension of such a structure, as has just been proved, is equal to 4. However, the hydrate shell may not always be correct. In these cases, the dimension of the hydration shell, together with the metal ion, can be greater than four. During the synthesis of organic compounds in nature, instead of water molecules, organic molecules can be located in the metal hydrate shell. An important role in nature is played by complex compounds around iron ions (the previously discussed Fe-porphyrins), which have a higher dimension. The chlorophyll molecule has a structure similar Fe-porphyrin. The only metal ion in chlorophyll is the magnesium ion. This structure also has a higher dimension.

10.5. Dimension and Genes

We simply understand that a gene is a certain segment of DNA responsible for the inheritance of a certain trait. However, this is not quite true. We know that DNA is located in chromosomes, but in addition to DNA, there are also proteins in chromosomes. This is not an inert part of the chromosomes since proteins with metal ions form enzymes. It is enzymes that organize all chemical processes in the body. In particular, they can cut DNA, highlighting certain sections from them. In addition, there are areas in DNA that are not responsible for the transmission of hereditary information – introns (Guttman, et al., 2002).

It has recently been experimentally established that protein histone modifications affect the active inheritance of traits, and this effect is not associated with the sequence of nitrogenous bases in nucleic acids (epigenetics) (Lindquist, et al., 2016; Sanbonmatsu, et al.,2016; Mancuso, 2017). It follows from this that the genes are more complex formations than just a segment of the sequence of nucleotides in DNA. Timofeeff-Ressovsky wrote about this in his theory of the target (Timofeeff, Zimmer, 1947). In particular Timofeeff-Ressovsky argued that the gene is a physics-chemical unit, i.e., a large molecule, micelle, or, in the broadest sense, a crystalline structure. Obviously, the dimension of such a unit will be much more than three, since even its constituent parts-nucleotides and proteins have a higher dimension. Timofeeff-Ressovsky studied the effect of irradiation on the processes of heredity. It should be noted here that these processes require additional analysis. The fact is that since a gene has a higher dimension, gene mutations under the influence of irradiation are the result of the movement of irradiating particles in a space of higher dimension. But in chapter 3 of this work, it is shown that even the inertial motions of particles in the space of higher dimension do not obey the laws of classical mechanics due to the non-Euclidean nature of the space of higher dimension (Zhizhin, 2021).

Conclusion

It has been proven that the molecules of living organisms have the highest dimension (more than three) (Zhizhin, 2016; 2019 b; 2020 c). This is confirmed by the data of calculating the dimensions of molecules of various carbohydrates, proteins, nucleic acids and other biomolecules. Water molecules, which are known to average 80% of the mass of any living organism, also have the highest dimension. Since it is believed that biomolecules were formed in the process of evolution from the LUKA protocell more than 4.5 billion years ago in layers of water on the surface of minerals (Qian, et al., Russel, et al., Simakov, 2018; Simakov, et al., 2020), apparently the higher dimension of biomolecules is a consequence of the higher dimension of water molecules. The highest dimension of water molecules is a consequence of the electronic structure of the atoms of the water molecule and the processes of hydration around the ions of mineral substances in water and on the surface of minerals. In this case, it should be taken into account that the elementary cells of minerals also have a higher dimension (Zhizhin, 2018) and water during hydration changes its structure complementary to the vicinity of mineral ions.

Summary

The geometry of living and non-living substances of nature on the planet Earth is considered. It has been convincingly proved that at the molecular level all objects of nature (living and non-living) have the highest dimension. Because of this, as shown, the geometry of all these objects is Euclidean (it does not hold the axioms of Euclidean geometry of any dimension). This, one might say, is earthly non-Euclidean. It is not related to the curvature of space, as is the case in hyperbolic and elliptic geometries related to possible outer spaces. Living and non-living parts of nature exist in the process of cooperation, carrying out the circulation of substances in nature. Living and non-living parts of nature have similar properties such as self-regulation and replication, and have structures of rather great complexity. We can say that living nature is a product of the development and self-regulation of inanimate nature.

It should be recognized that the designation of inanimate nature by the term inert matter, introduced 100 years ago by V.I. Vernadsky, is incorrect, since they do not reflect the complex processes of self-regulation in inanimate nature. Such a term contrasts both parts of nature, introducing an unnecessary antagonism that does not exist in nature.

References

Adam, N. K. 1938. *The Physics and Chemistry of surfaces.* Oxford: Clarendon Press.

Agol, V. I., Bogdanov, A. A., Gvozdev, V. A. 1990. *Molecular biology: Structure and biosynthesis of nucleic acids.* Higher School.

Akhmetov, S. F. 1990. *Faces of a Pomegranate*. Moscow: Science.

Albert, B. et al. 2002. *Molecular Biology of the Cell*, 4th edition. New York: Garland Science.

Alexandrov, A. D. 1956. Abstract space. *Mathematics, its content, methods and meaning. T.3.* Moscow: Publishing House of the Academy of Sciences of the USSR, 93 - 180.

Ashrafi, A. R., Cataldo, F., Iranmanesh, A., Ori, O. 2013. *Topological Modelling of Nanostructures and Extended Systems.* Springer.

Ashrafi, A. R. et al. 2013. Mathematics of D_5 Network. In M. V. Diudea, C. L. Nagy (Eds). *Diamond and Related Nanostructures*. Dordrecht, Heidelberg, New York, London: Springer. 321 – 334.

Audier, M., Pannetier, J., Leblanc, M., Janot, C., Lang, L. M., Dubost, B. 1988. An approach to the structure of quasicrystals: A single crystal X - Ray and neutron diffraction study of the R – phase. *Physics B,* 136 - 142.

Babenko, Yu. I. 2015. Antispace. In I. N. Toganov, Yu. I. Babenko. *Anti - time and antispace*. S.-Petersburg: TIN, 187 - 238.

Balaban, A. T. 2013. Diamond Hydrocarbons and Related Structures. In M.V. Diudea, C. L. Nagy (Eds). *Diamond and Related Nanostructures*. Dordrecht, Heidelberg, New York, London: Springer. 1 – 28.

Baumschliter, C. W. et al. 2007. Electronic and vibrational spectroscopy od diamondoids and the interstellar infrared bands between 3.35 and 3.55 μm. *Astrophys J. 671*:458 – 469.

Belov, N. V. 1976. *Essays on structural mineralogy*. Moscow: Nedra.

Beltrami, E. 1868. Staggio di Interpretazione della geometrica nonenclidea. *Giorm. Mat Napoli*. 6. Opera mat. I, 374 - 405.

Berg, J., Tymoczko, J. & Stryer, L. 2002. *Biochemistry*. San Francisco: W. H. Freeman & Co Ltd.

Bergman, G., Waugh, J. L. T., Pauling, L. 1952. Crystal structure of the intermetallic compound and related phases. *Nature* 169, 1057 - 1058.

Bergman, G., Waugh, J .L. T., Pauling, L. 1957. The crystal structure of the metallic phase. *Acta Cryst.* 10, 254 - 259.

Bogdanov, A. A. 1989. *Tectology. General Organizational Science.* Books 1, 2. Moscow: Economics.

Brachet, J. 1941 – 1942. La détection histochimique et la microdosage des acides pentosenucleiques. *Enzymologia,* 10, 87 – 96.

Bredley, A. J., Thewlis, J. 1926. The structure of γ - brass. *Proc. Roy. Soc. London A* 112, 678 - 681.

Buckour, I., Delanu, A. 1972. *Introduction to the theory of categories*. Moscow: World.

Bukhstaber, V. M., Panov, T. E. 2004. *Toric actions in topology and combinatorics*. Moscow: MCNMO Publishing House.

Butler, J. M. 2005. *Forensic DNA Typing*. London: Gardners Books.

Büchel, W. 1963. Warum hat ger Raume drei Dimensionen? *Physilalische Blatter, 19(12),* 547 -549.

Caly, L. et al. 2020. The FDA – approved Drug Invermectin inhibits the replication of SARS – CoV – 2 in vitro. *Elsevier Enhanced Reader.* 13.04.

Cantor, C. R., Schimmel, P. R. 1980. *Biophysical Chemistry. Part I. The Conformation of Biological Macromolecules.* San Francisco: W. H. Freeman & Co Ltd.

Cartan, E. 2010. *Geometry Riemann`s spaces*. Moscow: URSS.

Caspersson, T., Landeström – Hyden, H., Aquilonius, L. 1941. Cytoplasmsnukleotide in eiweissproduzierenden Drüsenzellen. *Chromosoma, 2,* 111 – 131.

Conway, J. H., Sloane, N. J. A. 1988. *Sphere Packings, Lattices, and Groups.* Springer – Verlag: New York.

Corey, R. B., Pauling, L. 1953. Fundamental dimensions of polypeptid chains. *Proc R Soc Lord B Biol Sci,* 141(902), 10-20.

Coxeter, H. S. M. 1963. *Regular Polytopes.* New York, London: John Wile & Sons, INC.Crick, F. H. C. 1959. The present position of the coding problem. *Brookhaven Symp. Biol.* 12: 35 – 39.

Crotty, S., Cameron, C., Audino, R. 2002. Rubavirin`s antiviral mechanism of action: lethal mutagenesis? Journal of Molecular Medicine, 80(2): 86 – 95.

Curie, P. 1908. *Qeuvres*. Paris.

Curie, P. 1966. On symmetry and physical phenomena: symmetry of electric and magnetic fields. *Classics of science. P. Curie. Selected Works*. Moscow - Leningrad: Science. 95 - 113.

Dahl, J. E., Liu, S. G., Carlson, R. M. K. 2003. Isolation and structures of higher diamondoids, nanometer – sized diamond molecules. *Science, 229*: 96 - 99 .

Dahm, R. 2005. Friedrich Miescher and the discovery of DNA. *Dev Biol, 278* (2): 274 – 288.

De Clerey, Em. 2004. Antiviral drugs in current clinical use. *Journal of Clinical Virology*. 30.2.

Delaunay, B. N. 1937. Geometry of positive quadratic forms. *Progresses of Mathematical Sciences,* 3, 16 - 62.

Delaunay, B. N. 1961. Proof of the main theorem of the theory of stereohedrons. *Reports of the Academy of Sciences of the USSR, 138 (6),* 1270 - 1272.

Delaunay, B., Sandakova, N. 1961. Theory of stereohedrons. *Proceedings of the Mathematical Institute. V. A. Steklov, 64*, 28 – 51.

Diudea, M. V., Nagy, C. L. 2007. *Periodic Nanostructures*. Springer.

Diudea, M. V. 2018. *Multi - shell Polyhedral Clusters*. Springer.

Dixon, M., Webb, E. C. *Ensymes*. 1979. Cambridge: Longman Group Ltd.

Eisenberg, D., Kautsman V. 1975. *Structure and properties of water*. Leningrad: Gidrometyoizdat.

Efimov N. V. 1945. *Higher geometry*. Moscow - Leningrad: OGIZ.

Eiji, A., Yanfa, Y., Pennycook, S. J. 2004. Quasicrystals as cluster aggregates. *Nature materials, 3*:759 - 767.

Einstein, A. 1930. Feld - und Ätherproblem in der Physik. *Die Koralle*, 5, 486 - 487.

Encyclopedia of Mathematics. 1977. Moscow: Publishing House Soviet Encyclopedia.

Ershov, Y. A. 1989. *Mechanisms of toxic action of inorganic compounds*. Moscow: Medicine.

Euclid. 2012. *Beginnings*. Moscow: URSS.

Fersmann, A. E. 1923. *Chemical elements of the Earth and Space*. Petrograd.

Fersmann, A. E. 1937. *Geochemistry. V.3*. Leningrad: ONTI.

Feyman, R. 1968. *The nature of physical laws*. Moscow: World. Fisher, J., Baumgartner, J., Marschner, C. 2005. Synthesis and structure of sila – adamantane. *Science, 310*: 825 - 830.

Fomenko, A. T. 1992. *Visual geometry and topology. Mathematical images in the real world*. Moscow: Moscow State University Publishing House.

Frank - Kamenetskiy, M. 1986. DNA structure: A simple solution to the stability of the double helix? *Nature*, *324*, 325 (27 November).

Frank - Kamenetskiy, M. 1988. H – form DNA and the hairpin – triplex model. *Nature*, *333*, 214 (19 May).

Frank – Kamenetskiy, M. D. 2004. *The DNA Age*. Moscow, Publishing house KDU.

Frank – Kamenetskiy, M. D. 2010. *Queen of a living cell. From the structure of DNA to the biotechnological revolution*. Moscow: Act – Press.

Freeman, I. M. 1969. Why is Space Three - Dimensional? *Am. J. Phys. 37,* 1222 - 1224.

Gamow, G. 1954. Possible relation between deoxyribonucleic acid and protein structures. *Nature,* 173, 318.

Gillespie, R. J. 1972. *Molecular Geometry*. New York, Cincinnati, Toronto, Melbourne: Van Nostrand Reinhold Company.

Gillespie, R. J., Hargittai, I. 1991. *The VSEPR Model of Molecular Geometry*. London : Allyn & Bacon.

Gosset, T. 1900. On the regular and semi - regular figures in space of n dimensions. *Messenger of Mathematics*. 29, 43 – 48.

Gratias, D., Cahn, J. W. 1986. Periodic and qusi – periodic crystals. *Scripta Metallurgica,* 20(9), 1193 – 1197.

Grünboum, B. 1967. *Convex Polytopes*. London: Springer.

Gubin, S.P. 1987. *Chemistry clusters. Basics of classification and structure*. Moscow: Science.

Gurevich, L., Mostepanenko, V. 1971. On the Existence of Atoms in *n* - Dimensional Space. *Phys. Letters A. 35(3)*, 201 - 202.

Ha, S. C. et al. 2005. Cristal structure of a junction between B – DNA and Z – DNA revels two extruded bases. *Nature, 437* (7062), 1183 – 1186.

Hawkes, E. J., et al. 2016. COOLAIR antisense RNAs from evolutionarily conserved elaborate secondary structures. *Cell Reports,* 16(12), 3087 – 3096.

Hegel, G. V. 1998. *The Science of Logic*. Moscow: Thought.

Hershey, A. & Chase, M. 1952. Independent functions of viral protein and nucleic acid in growth of bacteriophage. *J Gen Physiol, 36* (1): 39 – 56.

Hilbert, D., Cohn - Vossen, S. 1932. *Anschauliche Geometrie.* Berlin.

Hilbert, D. 1901. Gesammelte Abhandlagen. *Archivf. Mathematics and Phisics.* 2, 263-272.

Hobson, E. W. 1931. *The theory of spherical and ellipsoidal harmonics.* Cambridge: At the University Press.

Ivanov, K. P. 1993. *Fundamentals of the body's energy. Biological oxidation and oxygen supply*. St. Petersburg: Science.

Kadanoff, L. P. 1966. Scaling Laws for Ising Models Near. *Physics,* 2, 263 -272.

Karapet`yants, M. X., Drakin, S. I. 1994. *General and Inorganic Chemistry*. Moscow: Chemistry.

Klassen, V. I. 1973. *Water and magnet*. Moscow: Science.

Klein, F. 1928. *Vorlesungen Über Nicht - euklidische Geometrie*. Berlin: Springer.

Komura, Y., Sly, W. G., Shoemaker, D. P. 1960. The structure of the R phase, Mo - Cu - Cr. *Acta Cryst. 13*, 575 - 585.

Koolman, J., Roehm, K. 2013. *Color Atlas of Biochemistry.* Stutgart, New York: Thieme.

Kosrapohr, (Ed.). 2003. *Managing data mining technologies in organizations: techniques and applications.* Hershey, PA: Idea Group Publishing.

Kostromina, N. A., et al. 1990. *Chemistry of coordination compounds*. Moscow: Chemistry.

Kozlov, S. N., Stragunsky, L. S. 2002. *Modern antimicrobial chemotherapy. A guide for doctors*. Moscow: Borges.

Lalalyants, I. 1990. "Closure" in Paris. *Science and Life*, 1, 86 - 89.

Landa, S., Machacek, V. 1933. Sur l`adamantane, nouvel hydrocarbure extait du naphte. *Collection Czech Commun, 5* :1 – 5.

Landau, L. D. 1937. To Theory of Phase Transitions. *Journal of Experimental and Theoretical Physics*, **7**, 19 -38.

Landau, L. D., Lifshitz, E. M. 1963. *Quantum mechanics. Nonrelativistic theory.* Moscow: State Ed. Phys.- Math. Lit.

Lehn, J. M. 1998. *Supramolecular chemistry. Concept and perspectives.* Novosibirsk: Science.Lehninger, A. L. 1982. *Principes of Biochemistry.* Worth Publishers, Inc.

Lindquist, S. 2016. Luminidependens (LD) is an arabidopsis protein with prion behavior. *Proceedings of the National academy of sciences of the United States of America,* 113(21), 6065 - 6070.

Lobachevsky, N. I. 1829. About the beginnings of geometry. *Kazan University Bulletin.*

Lobachevsky, N. I. 1835. *Imaginary geometry*. Scientific notes of Kazan University.

Lord, E. A., Mackay, A. L., Ranganathan, S. 2006. *New Geometry for New Materials.* Cambridge: Cambridge University Press.

Lundberg, L., et al. 2013. Nuclear import and export inhibitors alter capsid protein distribution in mammalian cells and reduce Venezuelan Equine Encephalitis Virus replication. *Antiviral Res.* 100(3), 662-72.

Lyashchenko, A. K., Dunyashev, V. S. 2003. Complementary organization of the water structure. *Journal of Structural Chemistry,* 44 (5), 909 - 915.

Mackay, A. L. 1962. A dense non - crystalline packing of equal spheres. *Acta Cryst*. 15, 1916 - 1918.

Mancuso, S. 2017. *PLANT REVOLUTION. Le piante hanno già inventato il nostro futuro.* Firenze - Milano: GUIUTI EDITORE.

Martynova, O. I., et al. 1969. To the question of the mechanism of the influence of the magnetic field on aqueous solutions of salts. *Advances in physical sciences,* 98 (1), 195 - 199.

McMullen, P., Schulte, P. 2002. *Abstract Regular Polytopes*. Cambridge Univ. Press.

Metzler, D. E. 1977. *Biochemistry. The Chemical Reactions of Living Cells.*New York, San Francisco, London: Academic Press.

Miller, E. V. 1969. On the spontaneous decrease in the density of water condensed from steam. *Reports of the USSR Academy of Sciences,* 184 (1), 136 - 141.

Minkowski, H. 1911. Diskontinuitätshereich für arithemetische Äquivalenz. *Gesammelte Abhandlangen, Bd. II*, 70 - 72.

Mukhopadhyay, N. K. et al. 1993. Diffraction studies of icosahedral phases in. *Journal of Non - Crystalline Solids, 153 - 154*:1193 - 1197.

Nagy, C. L., Diudea, M. V. 2013. Diamond D_5. In M. V. Diudea, C. L. Nagy (Eds.). *Diamond and Related Nanostructures*. Dordrecht, Heidelberg, New York, London: Springer. 91 – 106.

Novikova, N. A. 2007. Storage and sale of virus genetic information. Lower Novgorod State University named N.I. Lobachevsky: Lower Novgorod.

Novikova, N. A., et al. 2002. *Virology*. Lower Novgorod State University named N.I. Lobachevsky: Lower Novgorod.

Nyman, H., Andersson S. 1979. On the structure of and γ - brass. *Acta Cryst. A* 35, 580 - 583.

Oswald, A. T., MacLeod, C. M., McCarty, M. 1944. Studies on the Chemical Nature of the Substance Inducing Transformation of Pneumococcal Types: Induction of Transformation by a Desoxyribonucleic Acid Fraction Isolated from Pneumococcus Type III. *Journal of Experimental Medicine, 79* (2): 137–158.

Paster, L. 1922. *Qeuvres*. 1. Paris.

Paster, L. 1960. *Selected words*. Moscow: Academy of Sciences of the USSR.

Pauling, L., Corey, R. B. 1950. Two Hydrogen - bonded Spiral Configurations of the Polypeptide Chain. *J. Am. Chem. Soc.*72, 5349.

Pauling, L., Corey, R. B., Branson, H. R. 1951. The Structure of Proteins: Two Hydrogen - bonded Helical Configurations of the Polypeptide Chain. *Proceedings of the national Academy of Sciences of the USA,* 37(4), 205 - 211.

Pauling, L. 1987. So - called Icosahedral Decagonal Quasicrystals are Twins of an 820 - atom Cubic Cristal. *Phys. Rev. Lett.* 58, 365-368.

Perrin, D. D., Armarego, W. L. F., Perrin, D. R. 1980. *Purification of Laboratory Chemical.* New York: Pergamon.

Poincare, A. 1895. Analysis situs. *J. de e`Ecole Polytechnique, 1,* 1 – 121.

Ponnuswamy, P., Gromiha, M. 1994. On the conformational stability of oligonucleotide duplexes and tRNA molecules. *J Theor Biol, 169*(4): 419 - 432.

Pontryagin, L. S. 1976. *The foundations of combinatorial topology.* Moscow: Science.

Pouling, L. 1960. *The Nature of the Chemical Bond and the Structure of Molecules and Crystals: An Introduction to Modern Structural Chemistry.* (3rd ed). Ithaca, New York: Cornell University Press.

Qian, G., et al. 2010. An experimental study of the mechanism of the replacement of magnetite by pyrite up to 300 ^{0}C. *Geochem. Cosmochim. Acta,* 74, 5610 - 5630.

Rahman, A., Stillinger, F. H. 1973. Hydrogen - Boden patterns in liquid water. *J. Am. Chem. Soc*. 95, 7943 - 7948.

Razonable, R. R. 2011. Antiviral Drugs for Viruses Other Than Humanimmunodeficiency virus. Mayo Clin Proc. 86(10): 1009 – 1026.

Riemann, B. 1854. *Ueber die Hypothesen, welche der Geometrie zu Grunde liegen.* University of Göttingen.

Robertson, S. A. 1984. *Polytopes and Symmetry.* Cambridge: Cambridge University Press.

Rodger, J. A. 2015. Discovery of medical Big Data analytics: Improving the prediction of traumatic brain injury survival rates by data mining Patient Informatics Processing Software Hybrid Hadoop Hive. *Informatics in Medicine Unlocked, 1,* 17 – 26.

Rodnikova, M. N. 1993. Features of solvents with a spatial network of H - bonds. *Journal of Physical Chemistry,* 67 (2), 275 - 280.

Russel, M. G., et al. 2005. On hydrothermal convection systems and the émergence of life. *Economic Geology*, 100, 419 - 438.

Samson, S. 1962. Crystal Strucrure of. *Nature*, 195, 259 - 262.

Samson, S. 1964. The crystal structures of semiconductors and a general valence rule. *Acta Cryst., 17, 49* - 51.

Samson, S. 1965. The crystal structure of the phase β. *Acta Cryst.* 19, 401 - 413.

Samson, S. *1967*. The Crystal Structure of the Intermetallic Compound*. *Acta Cryst. 23, 586 - 594.*

Samson, S. 1972. Complex cubic A_6B compounds. II. The crystal structure of Mg_6Pd. *Acta Cryst. B* 28, 936 - 945.

Samoilov, O. Y. 1957. *Structure of aqueous solutions and hydration of ions*. Moscow: Science.

Schrodinger, 1928. *Abhandlungen zur Wellenmechanik.* Leipzig.

Semenchenko, V. N. 1932. *What are solutions*. Moscow - Leningrad: Science.

Shechtman, D., et al. 1984. Metallic phase with longerange orientational order and no translational symmetry. *Phys. Rev. Lett.* 53, 1951 - 1953.

Shevchenko, V., Zhizhin. G., Mackay, A. 2013. On the structure of the quasicrystals in the high dimension space. In M. V. Diudea, C. L. Nagy (Eds.). *Diamond and Related Nanostructures*. Dordrecht, Heidelberg, New York, London: Springer. 311 – 320.

Shevchenko, V. Y. 2011. *Search in Chemistry, Biology and Physics of the Nanostate.* Saint - Petersburg: Lema.

Shevchenko, V., Zhizhin, G., Mackay, A. 2013 b. On the structure of the quasicrystals in the high dimension space. *News RAS. Chemical series*, 2, 269 - 274.

Simakov, S., et al. 2018. Nano - and micro - size diamond genesis in nature: an overview. *Geoscience Frontiers*, 9, 1849 - 1858.

Simakov, S., et al. 2020. Sicilian serpentinite xenoliths containing abiotic organics with nanodiamond clusters as key model for prebiotic processes. *Geoscience Frontiers*, https://doi.org/10.1016 /j.gsf.2020.04.008.

Sokolov M. S. 2013. V. I. Vernadsky and his biospherology. Biosphere, 5(3), 9 - 20.

Spirin, A. S. 1986. *Molecular biology. Structure of the ribosome and protein biosynthesis.* Moscow: High school.

Spirin, A. S., Gavrilova, L. P. 1971. *The Ribosome*. Moscow: Science.

Steed, J. V., Atwood, J. L. 2000. *Supramolecular chemistry.* Chichester, UK: John Wiley and Sons.

Stewart, M., McLachlan, A. D. 1975. *Nature, 257*, 331 – 333.

Tatarsky, F. 1989. Struggle for "strange" water. *Chemistry and Life,* 12, 57 - 59.

Timofeeff – Ressovsky, N. W., Zimmer, K. G. 1947. *BIOPHYSIK. Das Trefferprinzip in der Biologie.* Leipzig: Arbeitsgemeinschaft medizinischer Verlage G.m.b.H.

Vargaftik, M. N., et al., 1985. *Reports of the Academy of Sciences of the USSR.* 284, 4, 896-899.

Venkov, B. A. 1959. On the design of parallelohedrons. *Mathematical collection.* 49 (2), 207 -224.

Vernadsky, V. I. 1917 – 1921. *Diaries.* Kiev: Naukova Thought.

Vernadsky, V. 1925. *History of the minerals of the earth's crust.* Leningrad.

Vernadsky, V. I. 1926. *The Biosphere*. Springer.

Vernadsky, V. I. 1977. *Reflections of a naturalist. Scientific thought as a planetary phenomenon.* Moscow: Science.

Vilenkin, N. V. 1969. *Combinatorics.* Moscow: Science.

Watson, J. D., Crick, F. H. C. 1953a. Molecular structure of nucleic acids. *Nature, 171*, 738 – 740.

Watson, J. D., Crick, F. H. C. 1953b. Genetical implications of the structure of deoxyribose nucleic acid. *Nature, 171*, 964 - 967.

Whitehead, A. N. 1990. *Selected works on philosophy*. Moscow: Progress.

Wulan, W. N., et al. 2015. Nucleocytoplasmic transport of nucleocapsid proteins of enveloped RNA viruses. *Front Microbiol,* 6, 553.

Yanshin, F.T. 2011. Significance of the doctrine of the biosphere in the work of V.I. Vernadsky. Biosphere, 3(1), 18 - 26.

Zang, X., Kelton, K. 1993. High - order crystal approximant alloys $Ti_{54}Zr_{26}Ni_{20}$. *Journal of Non - Crystalline Solids,* 153-154: 114 - 118.

Zhizhin, G. V. 2005. *The Dissipative Structures in Chemical, Geological, and Ecological Systems*. S. - Petersburg: Science.

Zhizhin, G. V. 2010. *Geometrical bases of the dissipative structures*. S. Petersburg: Polytechnica.

Zhizhin, G. V. 2012. Hierarchical filling of spaces with polytopes. *In Science and human progress. 7th St. Petersburg meeting of Nobel Prize Winners.* St. Petersburg: Publishing Hause of Polytechnic University.

Zhizhin, G. V. 2013 a. Images of convex regular and semiregular *n* - dimensional polytopes. *Proceedings of the 9th All - Russian Scientific School "Mathematical Research in the Natural Sciences."* Apatite: Geological Institute, KSC RAS. 32 - 42.

Zhizhin, G. V. 2013 b. Relations for the number of faces of different dimensions in the tower of n - dimensional convex polytopes. *Materials 9th All - Russian Scientific School "Mathematical research in the natural sciences."* 24 - 32. Geological institute KSC RAS, Apatity.

Zhizhin, G. V. 2014 a. On Higher Dimensionalities in Nature. *Biosphere*, 6(4), 313 - 318.

Zhizhin, G. V. 2014 b. *World – 4D.* St. – Petersburg: Polytechnic Service.

Zhizhin, G. V. 2014 c. Fractal nature of disproportionate phases. *Proceedings of the109th All-Russian Scientific School "Mathematical Research in the Natural Sciences."* Apatity, Geological Institute KSC RAS: 9 - 14.

Zhizhin, G. V. 2014 d. Incommensurable and fluctuating structures in the terrestrial space. *Biosphere, 3*: 211 - 221.

Zhizhin, G. V. 2015 a, November. *Polytopic prismahedrons – fundamental regions of the n -dimension nanostructures.* Paper presented at The International conference "Nanoscience in Chemistry, Physics, Biology and Mathematics," Cluj – Napoca, Romania.

Zhizhin, G. V. 2015 b. Dimensions of compounds in supramolecular chemistry. *Int. J. Chem. Modeling* 7(2), 215 – 238.

Zhizhin, G. V. 2015 c. The dimensions of supramolecular compounds. *Biosphere,* 7(2), 149 – 154.

Zhizhin, G. V. 2016 a. The structures, topological and functional dimension of biomolecules. *Int. Journal of Chemoinformatic and Chemical Engineering,* 5(2), 44 - 58.

Zhizhin, G. V. 2016 b. N – prismahedrons and their dual polytopes. *International Journal Chemical Modeling 5(4).*

Zhizhin, G. V. 2016 c. From multidimensional phase spaces to multidimensionality of space in Nature. *Biosphere* 8(3), 258 - 267.

Zhizhin, G. V., Diudea, M. V. 2016. Space of nanoworld. In M.V. Putz, C.M. Marius (Eds.), *Sustainable nanosystems, development, properties, and applications (pp. 221 - 245),* Hershey PA, USA: IGI Dlobal.

Zhizhin G. V., Khalaj Z., Diudea M. V. 2016. Geometrical and topology dimensions of the diamond. In A. R. Ashrafi, M. V. Diudea (Eds.). *Distance, symmetry, and topology in carbon nanomaterials (*pp. 167- 188*).* New York, London: Springer.

Zhizhin, G.V. 2018. *Chemical Compound Structures and the Higher Dimension of Molecules: Emerging Research and Opportunities.* Hershey PA, USA: IGI Global.

Zhizhin, G. V. 2019 a. *The Geometry of Higher - Dimensional Polytopes*, IGI Global.

Zhizhin, G. V. 2019 b. *Attractors and higher dimensions in population and molecular biology: emerging research and opportunities*. Hershey PA: IGI Global.

Zhizhin, G. V. 2019 c. The Structure of the Polytope of Hereditary Information. *In. J. of Chemoinformatics and chemical Engineering.* 8(2), 7 - 22.

Zhizhin, G. V. 2019 d. Higher Dimensions of Clusters of Intermetallic Compounds. *International Journal of Applied Nanotechnolodgy Research,* 4(1), 8 - 25.

Zhizhin, G. V. 2019 e. The Geometry of Higher – Dimensional Multi – Shell Clusters with Common Center and Different Center: The Geometry of Metal Clusters with Ligands. *Int. Journal of Applied Nanotechnology Research,* 4(2), 45 - 65.

Zhizhin, G. V. 2019 f. About the True Equilibrium in the Population of the Plants. Journal of Biochemistry and Molecular Biology Research, 4(1), 214 – 224.

Zhizhin, G. V. 2019 g. The Metric Structure of Nucleic Acids and Higher Dimension of Their Constituents. *J. Chemoinformatics and Chemical Engineering, 7(2),* 1 – 15.

Zhizhin, G. V. 2019 h. The Law of Conservation of Incidents in the Space of Nanoworld. *J. Chemoinformatics and Chemical Engineering.* 8(1), 1 – 15.

Zhizhin, G. V. 2020 a. The Dimension and Life. *Acta Scientific Biotechnology* ,1(11), 35 - 40.

Zhizhin, G. V. 2020 b. Spirals of Polysaccharides Simultaneously Rotating in Opposite Directions. *Acta Scientific Biotechnology,* 1(7), 13 - 16.

Zhizhin, G. V. 2020 c. Hidden Nucleic Acids Bond Order? *Acta Scientific Biotechnology* 1(4): 34 -36.

Zhizhin, G. V. 2020 d. Interaction of Nucleic Acids: Hidden Order of Interaction. *International Journal of Applied Research in Bioinformatics,* 11(2): 1 -15.

Zhizhin, G. V. 2020 e. Higher Dimensions in Linear, Helicoidal and Folder Structures of Protein Molecules. *Acta Scientific Biotechnology* ,1(12), 02 - 09.

Zhizhin, G. V. 2021 a. Hidden Order of Interaction Nucleic Acids. *Journal of Nature Science and Sustainable Technology,* 15(4): 1 – 5.

Zhizhin, G. V. 2021 b. Spatial Models of Sugars and their Compounds. *International Journal of Applied Research in Bioinformatics,* 11(2): 9 -23.

Zhizhin, G. V. 2021 c. *Nanotechnology and Clusters in the Space of Higher Dimension: Emerging Research and Opportunities:* IGI Global.

Zhizhin, G. V. 2021 d. *Normal Partitions and Hierarchical Filling of N - Dimensional Spaces:* IGI Global.

Zhizhin, G. V. 2021 e. The Structures of DNA Taking into Account the Higher Dimensional of its Components. *In Kosrupohr (Ed.) Encyclopedia of Organizational Knowledge, Administration, and Technologies.* Hershey, PA: IGI Global.

Zhizhin, G. V. 2021 f. Methylated Polytope of Hereditary Information. *In. J. of Applied Research on Public Health Management.* 6(2), 49 - 60.

Zhizhin, G. V. 2021 g. Native Structure of Globular Proteins in the Space of the Highest Dimension. *Journal of Nature Science and Sustainable Technology.* 15(3), 87 - 96.

Zhizhin, G. V. 2021 h. Features of Inertial Motion in the Space of the Highest Dimension. *Journal of Nature Science and Sustainable Technology*, 15 (4), 1 – 9.

Zhizhin, G. V. 2021 j. Spatial Structures of Fibrillar Proteins. *Int. J. of Applied Research in Bioinformatics.* 12(1).

Zhizhin, G. V. 2022 a. On the Possible Spatial Structures of the β – Amyloid. Towards the theory of the native structure of proteins. *Int. J. of Applied Research on Public Health Management,* 20, Jan.

Zhizhin, G.V. 2022 b. *The Classes of Polytopes Higher Dimensions in Physics, Chemistry and Biology.* Hershey PA, USA: IGI Global.

Ziegler, G. 1995. *Lectures on polytopes*. New York: Springer.

Zundel, G. 1972. *Hydrotation and intermolecular interactions*. Moscow: Mir.

Zwiebach, B. 2009. *A first course in string theory.* Massachusetts Institute of Technology.

About the Author

Gennadiy Zhizhin, PhD
Professor
Academician Russian Academy of Natural Sciences (Russia)
ORCID: 0000-0002-4933-7590
Email: genzhizhin@gmail.com

Dr. Gennadiy Zhizhin is the Head of the Russian Federation Branch of European Society of Mathematical Chemistry (2017 -); Chief Research of the OOO "Adamant", Skolkovo, Moscow, Russia (2017-2020); Senior Fellow of the Institute of Silicate Chemistry, Russian Academy of Sciences, St. Petersburg, Russia (2012-2014); Professor at North-Western State Technical University, Department of Chemistry, Computational Mathematics, Computer Science, Applied Mathematics, St. Petersburg, Russia (1994-2012); Professor of the Department of Mathematics of the National University of the mineral resource (Mining University), St. Petersburg, Russia (2012-2013); Academic degree of Doctor of Technical Science (1991); Academic rank of Professor (1999); Corresponding Member of the Russian Ecological Academy (2006-); Academician of the Russian Academy of Natural Sciences (2017-); Member of the Scientific Council on Combustion and Explosion of the Russian Academy of Sciences (2010-); Member of the editorial board of the journal *Biosphere* (2008-); Deputy editor-in chief of the *IJARB* (2020-); Member of the editorial board *IJCCE* (2019-2020).

Index

A

adamantane, 69, 70, 71, 72, 73, 74, 75, 76, 77, 78, 79, 80, 81, 82, 269, 270
amino acid, 156, 159, 160, 161, 162, 164, 165, 166, 167, 169, 170, 171, 173, 174, 175, 176, 177, 178, 179, 180, 186, 187, 212, 220, 231, 232, 251, 256, 259, 261
amyloid, 159, 187, 275
atom, 40, 53, 66, 67, 78, 80, 81, 82, 83, 84, 85, 86, 89, 92, 93, 110, 111, 112, 114, 115, 116, 125, 128, 131, 133, 135, 138, 141, 145, 149, 150, 151, 152, 154, 155, 156, 161, 166, 177, 183, 185, 189, 193, 194, 196, 198, 199, 200, 205, 206, 209, 210, 212, 213,214, 216, 217, 246, 254, 255, 257, 258, 261, 271
atomic orbital, 149, 253
atomic orbitals, 149

B

biomolecules, vii, ix, 101, 127, 191, 246, 253, 256, 263, 274
biopolymer, 190
biosphere, vii, ix, 273
biosynthesis, 189, 219, 267, 273
bonds, 71, 78, 80, 90, 92, 95, 127, 134, 142, 149, 150, 152, 154, 155, 156, 159, 162, 165, 168, 169, 174, 177, 183, 190, 196, 201, 203, 206, 207, 208, 209, 210, 211, 214, 216, 217, 219, 220, 240, 247, 255, 256, 261, 272

C

chain, viii, 1, 82, 127, 128, 130, 131, 135, 136, 140, 144, 147, 155, 156, 157, 159, 161, 164, 165, 166, 167, 169, 178, 187, 190, 192, 196, 197, 198, 200, 205, 206, 207, 208, 212, 213, 216, 220, 226, 228, 256
chemical bonds, 83, 89, 104, 105, 109, 125, 131, 134, 148, 165, 183, 189, 194, 201, 208, 214, 216, 229, 231
covalent bond, 80, 128, 149, 160, 162, 169, 176, 183, 189, 214, 247
cross-polytope, 10, 15, 16, 24, 25, 28, 29, 36, 37, 50, 51, 75, 79, 90, 91, 92, 103, 107, 116, 119, 120, 121, 122, 123, 124, 125, 126, 159, 178, 179, 182, 219, 224, 225, 227, 228, 229, 234, 235, 236, 237, 238, 239, 247, 249, 250

D

dimension of the space, 52, 59, 66, 219
dimensionality, viii, ix, 39, 44, 53, 54, 67, 69, 75, 79, 80, 82, 93, 126, 132, 164, 167, 191, 192, 200, 217, 218, 223, 225, 227, 233, 251

E

electron pairs, 67, 80, 92, 93, 109, 200, 214, 254
elliptic geometry, 1, 5
Euclidean geometry, v, viii, 1, 2, 3, 7, 10, 40, 45, 67, 69, 265
Euclidean space, vii, 5, 7, 8, 35, 40, 41, 42, 46

F

formation, 36, 69, 75, 77, 78, 80, 100, 103, 109, 110, 119, 123, 125, 126, 127, 128, 149, 154, 157, 158, 160, 161, 166, 169, 174, 177, 187, 189, 191, 192, 198, 214, 217, 220, 221, 225, 226, 235, 246, 256, 259, 261

G

gene(s), 42, 219, 239, 246, 251, 256, 262
genetic code, viii, 221, 231
genetic information, 189, 218, 219, 231, 240, 271
geometry, vii, viii, ix, 1, 2, 3, 4, 5, 6, 7, 8, 9, 10, 29, 39, 41, 42, 43, 44, 67, 68, 69, 127, 163, 166, 185, 191, 192, 205, 219, 233, 250, 251, 265, 269, 270

H

hereditary, v, viii, 189, 219, 220, 225, 228, 229, 230, 231, 232, 233, 239, 240, 241, 247, 248, 249, 250, 251, 256, 262, 274, 275
higher-dimensional space, 1, 3, 9, 39, 67, 69, 127, 160
hydrogen atoms, 70, 71, 128, 133, 136, 141, 163, 199, 201, 254, 255
hydrogen bonds, 159, 169, 174, 177, 190, 208, 219, 220, 255, 256, 261
hydroxyl groups, 82, 128, 131, 136, 141, 154, 193, 198, 199, 244
hyperbolic geometry, 1, 4

I

incidence, 9, 10, 11, 12, 13, 14, 18, 20, 21, 23, 24, 25, 26, 28, 29, 31, 35, 36, 37, 39, 40, 43, 48, 49, 50, 51, 52, 67, 69, 134, 219, 233, 234, 235, 236, 237, 239, 240, 249, 251
incidence coefficient, 12, 13, 14, 29, 35, 36, 48, 49, 51, 134, 219, 233, 234, 235, 236, 237, 249
inertial motion, 39, 40, 46, 52, 67, 263, 275
initial state, 125, 175, 217, 224
intermetallic compounds, 104, 232
intermolecular interactions, 275
introns, 262
isomer(s), , 127, 128, 135, 136, 148, 149

K

KBr, 85

L

linear molecules, 80, 140, 147, 157

M

memory of water, 253, 257
molecule, viii, 70, 71, 72, 80, 81, 82, 86, 89, 90, 103, 105, 106, 109, 110, 125, 128, 130, 131, 133, 134, 135, 136, 137, 138, 143, 144, 145, 148, 149, 150, 152, 153, 154, 155, 156, 157, 161, 162, 164, 166, 167, 169, 175, 176, 177, 178, 186, 189, 190, 191, 193, 195, 196, 197, 198, 199, 200, 201, 202, 203, 204, 205, 206, 207, 208, 210, 211, 212, 213, 214, 216, 217, 218, 220, 221, 225, 226, 231, 240, 254, 256, 261, 262, 263
monosaccharide, 127, 135, 138, 139, 140, 141, 142, 143, 145, 146, 147, 148, 149, 153, 154, 157, 158, 256

N

n-cross-polytope, 9, 10, 13, 14, 15, 35, 36, 51, 219, 235, 237, 238, 239
nitrogen bases, 156, 190, 219, 228, 241, 246, 247, 250
non-Euclidean geometry, v, ix, 1, 4, 6, 7, 8, 9, 39, 44, 45, 69
nucleic acid, viii, 127, 156, 189, 190, 191, 198, 200, 203, 204, 205, 208, 210, 211, 212, 213, 214, 215, 216, 217, 218, 219, 220, 221, 226, 227, 228, 230, 232, 241, 244, 246, 247, 250, 251, 256, 262, 263, 267, 270, 273
nucleotides, viii, 156, 190, 208, 209, 210, 216, 220, 232, 241, 246, 251, 262